TECHNOLOGY, GENDER AND HISTORY IN IMPERIAL CHINA

What can the history of technology contribute to our understanding of late imperial China? Most stories about technology in pre-modern China follow a well-worn plot: in about 1400 after an early ferment of creativity that made it the most technologically sophisticated civilization in the world, China entered an era of technical lethargy and decline. But how are we to reconcile this tale, which portrays China in the Ming and Qing dynasties as a dying giant that had outgrown its own strength, with the wealth of counterevidence affirming that the country remained rich, vigorous and powerful at least until the end of the eighteenth century? Does this seeming contradiction mean that the stagnation story is simply wrong, or perhaps that technology was irrelevant to how imperial society worked? Or does it imply that historians of technology should ask better questions about what technology was, what it did and what it meant in pre-modern societies such as late imperial China?

In this book, Francesca Bray explores subjects such as technology and ethics, technology and gendered subjectivities (both female and male), and technology and statecraft to illuminate how material settings and practices shaped topographies of everyday experience and ideologies of government, techniques of the self and technologies of the subject. Examining technologies ranging from ploughing and weaving to drawing pictures, building a house, prescribing medicine or composing a text, this book offers a rich insight into the interplay between the micro- and macro-politics of everyday life and the workings of governmentality in late imperial China, showing that gender principles were woven into the very fabric of empire, from cosmology and ideologies of rule to the material foundations of the state and the everyday practices of the domestic sphere.

This authoritative text will be welcomed by students and scholars of Chinese history, as well as those working on global history and the histories of gender, technology and agriculture. Furthermore, it will be of great use to those interested in social and cultural anthropology and material culture.

Francesca Bray is Professor of Social Anthropology at the University of Edinburgh, UK.

Asia's Transformations

Edited by Mark Selden, Cornell University, USA

The books in this series explore the political, social, economic and cultural consequences of Asia's transformations in the twentieth and twenty-first centuries. The series emphasizes the tumultuous interplay of local, national, regional and global forces as Asia bids to become the hub of the world economy. While focusing on the contemporary, it also looks back to analyze the antecedents of Asia's contested rise.

This series comprises several strands:

Asia's Transformations

Titles include:

1. Debating Human Rights*
Critical essays from the United States and Asia
Edited by Peter Van Ness

2. Hong Kong's History*
State and society under colonial rule
Edited by Tak-Wing Ngo

3. Japan's Comfort Women*
Sexual slavery and prostitution during World War II and the US occupation
Yuki Tanaka

4. Opium, Empire and the Global Political Economy*
Carl A. Trocki

5. Chinese Society*
Change, conflict and resistance
Edited by Elizabeth J. Perry and Mark Selden

6. Mao's Children in the New China*
Voices from the Red Guard generation
Yarong Jiang and David Ashley

7. Remaking the Chinese State*
Strategies, society and security
Edited by Chien-min Chao and Bruce J. Dickson

8. Korean Society*
Civil society, democracy and the state
Edited by Charles K. Armstrong

9. The Making of Modern Korea*
Adrian Buzo

10. The Resurgence of East Asia*
500, 150 and 50 year perspectives
Edited by Giovanni Arrighi, Takeshi Hamashita and Mark Selden

11. Chinese Society, second edition*
Change, conflict and resistance
Edited by Elizabeth J. Perry and Mark Selden

12. Ethnicity in Asia*
Edited by Colin Mackerras

13. The Battle for Asia*
From decolonization to globalization
Mark T. Berger

14. State and Society in 21st Century China*
Edited by Peter Hays Gries and Stanley Rosen

15. Japan's Quiet Transformation*
Social change and civil society in the 21st century
Jeff Kingston

16. Confronting the Bush Doctrine*
Critical views from the Asia-Pacific
Edited by Mel Gurtov and Peter Van Ness

17. China in War and Revolution, 18951949*
Peter Zarrow

18. The Future of US–Korean Relations*
The imbalance of power
Edited by John Feffer

19. Working in China*
Ethnographies of labor and workplace transformations
Edited by Ching Kwan Lee

20. Korean Society, second edition*
Civil society, democracy and the state
Edited by Charles K. Armstrong

21. Singapore*
The state and the culture of excess
Souchou Yao

22. Pan-Asianism in Modern Japanese History*
Colonialism, regionalism and borders
Edited by Sven Saaler and J. Victor Koschmann

23. The Making of Modern Korea, 2nd Edition*
Adrian Buzo

24. Re-writing Culture in Taiwan*
Edited by Fang-long Shih, Stuart Thompson and Paul-François Tremlett

25. Reclaiming Chinese Society*
The new social activism
Edited by You-tien Hsing and Ching Kwan Lee

26. Girl Reading Girl in Japan*
Edited by Tomoko Aoyama and Barbara Hartley

27. Chinese Politics*
State, society and the market
Edited by Peter Hays Gries and Stanley Rosen

28. Chinese Society, third edition*
Change, conflict and resistance
Edited by Elizabeth J. Perry and Mark Selden

29. Mapping Modernity in Shanghai
Space, gender, and visual culture in the Sojourners' City, 1853–98
Samuel Y. Liang

30. Minorities and Multiculturalism in Japanese Education
An interactive perspective
Edited by Ryoko Tsuneyoshi, Kaori H Okano and Sarane Boocock

31. Japan's Wartime Medical Atrocities
Comparative inquiries in science, history, and ethics
Edited by Jing-Bao Nie, Nanyan Guo, Mark Selden and Arthur Kleinman

32. State and Society in Modern Rangoon
Donald M. Seekins

33. Learning Chinese, Turning Chinese*
Becoming sinophone in a globalised world
Edward McDonald

34. Aesthetic Constructions of Korean Nationalism
Spectacle, politics and history
Hong Kal

35. Popular Culture and the State in East and Southeast Asia
Edited by Nissim Otmazgin and Eyal Ben Ari

36. Japan's Outcaste Abolition
The struggle for national inclusion and the making of the modern state
Noah Y. McCormack

37. The Market and Temple Fairs of Rural China
Red Fire
Gene Cooper

38. The Role of American NGOs in China's Modernization
Invited influence
Norton Wheeler

39. State, Society and the Market in Contemporary Vietnam
Property, power and values
Edited by Hue-Tam Ho Tai and Mark Sidel

40. East Asia beyond the History Wars
Confronting the Ghosts of Violence
Tessa Morris-Suzuki, Morris Low, Leonid Petrov and Timothy Yun Hui Tsu

Asia's Great Cities

Each volume aims to capture the heartbeat of the contemporary city from multiple perspectives emblematic of the authors' own deep familiarity with the distinctive faces of the city, its history, society, culture, politics and economics, and its evolving position in national, regional and global frameworks. While most volumes emphasize urban developments since the Second World War, some pay close attention to the legacy of the longue durée in shaping the contemporary. Thematic and comparative volumes address such themes as urbanization, economic and financial linkages, architecture and space, wealth and power, gendered relationships, planning and anarchy, and ethnographies in national and regional perspective. Titles include:

1. Bangkok*
Place, practice and representation
Marc Askew

2. Representing Calcutta*
Modernity, nationalism and the colonial uncanny
Swati Chattopadhyay

3. Singapore*
Wealth, power and the culture of control
Carl A. Trocki

4. The City in South Asia
James Heitzman

5. Global Shanghai, 1850–2010*
A history in fragments
Jeffrey N. Wasserstrom

6. Hong Kong*
Becoming a global city
Stephen Chiu and Tai-Lok Lui

Asia.com

Asia.com is a series that focuses on the ways in which new information and communication technologies are influencing politics, society and culture in Asia. Titles include:

1. Japanese Cybercultures*
Edited by Mark McLelland and Nanette Gottlieb

2. Asia.com*
Asia encounters the Internet
Edited by K. C. Ho, Randolph Kluver and Kenneth C. C. Yang

3. The Internet in Indonesia's New Democracy*
David T. Hill and Krishna Sen

4. Chinese Cyberspaces*
Technological changes and political effects
Edited by Jens Damm and Simona Thomas

5. Mobile Media in the Asia-Pacific
Gender and the art of being mobile
Larissa Hjorth

6. Online@AsiaPacific
Mobile, Social and Locative Media in the Asia–Pacific
Larissa Hjorth and Michael Arnold

Literature and Society

Literature and Society is a series that seeks to demonstrate the ways in which Asian literature is influenced by the politics, society and culture in which it is produced. Titles include:

1. The Body in Postwar Japanese Fiction
Douglas N. Slaymaker

2. Chinese Women Writers and the Feminist Imagination, 1905–1948*
Haiping Yan

Routledge Studies in Asia's Transformations

Routledge Studies in Asia's Transformations is a forum for innovative new research intended for a high-level specialist readership. Titles include:

1. The American Occupation of Japan and Okinawa*
Literature and memory
Michael Molasky

2. Koreans in Japan*
Critical voices from the margin
Edited by Sonia Ryang

3. Internationalizing the Pacific
The United States, Japan and the Institute of Pacific Relations in war and peace, 1919–1945
Tomoko Akami

4. Imperialism in South East Asia*
'A fleeting, passing phase'
Nicholas Tarling

5. Chinese Media, Global Contexts*
Edited by Chin-Chuan Lee

6. Remaking Citizenship in Hong Kong*
Community, nation and the global city
Edited by Agnes S. Ku and Ngai Pun

7. Japanese Industrial Governance
Protectionism and the licensing state
Yul Sohn

8. Developmental Dilemmas*
Land reform and institutional change in China
Edited by Peter Ho

9. Genders, Transgenders and Sexualities in Japan*
Edited by Mark McLelland and Romit Dasgupta

10. Fertility, Family Planning and Population Policy in China*
Edited by Dudley L. Poston, Che-Fu Lee, Chiung-Fang Chang, Sherry L. McKibben and Carol S. Walther

11. Japanese Diasporas*
Unsung pasts, conflicting presents and uncertain futures
Edited by Nobuko Adachi

12. How China Works*
Perspectives on the twentieth-century industrial workplace
Edited by Jacob Eyferth

13. Remolding and Resistance among Writers of the Chinese Prison Camp
Disciplined and published
Edited by Philip F. Williams and Yenna Wu

14. Popular Culture, Globalization and Japan*
Edited by Matthew Allen and Rumi Sakamoto

15. medi@sia*
Global media/tion in and out of context
Edited by Todd Joseph Miles Holden and Timothy J. Scrase

16. Vientiane*
Transformations of a Lao landscape
Marc Askew, William S. Logan and Colin Long

17. State Formation and Radical Democracy in India
Manali Desai

18. Democracy in Occupied Japan*
The U.S. occupation and Japanese politics and society
Edited by Mark E. Caprio and Yoneyuki Sugita

19. Globalization, Culture and Society in Laos*
Boike Rehbein

20. Transcultural Japan*
At the borderlands of race, gender, and identity
Edited by David Blake Willis and Stephen Murphy-Shigematsu

21. Post-Conflict Heritage, Post-Colonial Tourism
Culture, politics and development at Angkor
Tim Winter

22. Education and Reform in China*
Emily Hannum and Albert Park

23. Writing Okinawa: Narrative Acts of Identity and Resistance
Davinder L. Bhowmik

24. Maid in China*
Media, Morality, and the Cultural Politics of Boundaries
Wanning Sun

25. Northern Territories, Asia-Pacific Regional Conflicts and the Åland Experience
Untying the Kurillian knot
Edited by Kimie Hara and Geoffrey Jukes

26. Reconciling Indonesia
Grassroots agency for peace
Birgit Bräuchler

27. Singapore in the Malay World*
Building and breaching regional bridges
Lily Zubaidah Rahim

28. Pirate Modernity*
Delhi's media urbanism
Ravi Sundaram

29. The World Bank and the post-Washington Consensus in Vietnam and Indonesia
Inheritance of loss
Susan Engel

30. China on Video
Smaller Screen Realities
Paola Voci

31. Overseas Chinese, Ethnic Minorities and Nationalism
De-Centering China
Elena Barabantseva

Critical Asian Scholarship

Critical Asian Scholarship is a series intended to showcase the most important individual contributions to scholarship in Asian studies. Each of the volumes presents a leading Asian scholar addressing themes that are central to his or her most significant and lasting contribution to Asian studies. The series is committed to the rich variety of research and writing on Asia, and is not restricted to any particular discipline, theoretical approach or geographical expertise.

1. Southeast Asia*
A testament
George McT. Kahin

2. Women and the Family in Chinese History*
Patricia Buckley Ebrey

3. China Unbound*
Evolving perspectives on the Chinese past
Paul A. Cohen

4. China's Past, China's Future*
Energy, food, environment
Vaclav Smil

5. The Chinese State in Ming Society*
Timothy Brook

6. China, East Asia and the Global Economy*
Regional and historical perspectives
Takeshi Hamashita
Edited by Mark Selden and Linda Grove

7. The Global and Regional in China's Nation-Formation*
Prasenjit Duara

8. Decoding Subaltern Politics*
Ideology, disguise, and resistance in agrarian politics
James C. Scott

9. Mapping China and Managing the World*
Culture, cartography and cosmology in late Imperial times
Richard J. Smith

10. Technology, Gender and History in Imperial China*
Great transformations reconsidered
Francesca Bray

* Available in paperback

TECHNOLOGY, GENDER AND HISTORY IN IMPERIAL CHINA

Great transformations reconsidered

Francesca Bray

Routledge
Taylor & Francis Group
LONDON AND NEW YORK

First published 2013
by Routledge
4 Park Square, Milton Park, Abingdon, Oxon OX14 4RN
605 Third Avenue, New York, NY 10017

Routledge is an imprint of the Taylor & Francis Group, an informa business

British Library Cataloguing in Publication Data
A catalogue record for this book is available from the British Library

Library of Congress Cataloging in Publication Data
Bray, Francesca.
Technology, gender and history in imperial China: great transformations reconsidered/Francesca Bray.
p. cm. – (Asia's transformations)
Includes bibliographical references and index.
1. Technological innovations – China – History. 2. Technology – Social aspects – China. 3. Sex role – China – History 4. China – Social conditions – 960–1644. I. Title.
T27.C5B69 2013
303.48'30820951 – dc23 2012043561

ISBN: 978-0-415-63956-9 (hbk)
ISBN: 978-0-415-63959-0 (pbk)
ISBN: 978-0-203-08330-7 (ebk)

Typeset in Bembo
by Florence Production Ltd, Stoodleigh, Devon, UK

CONTENTS

FIGURES

ACKNOWLEDGEMENTS

My deepest gratitude to Mark Selden for inviting me to contribute a volume to the series *Asia's Transformations/Critical Asian Scholarship*, for constant encouragement over the years it took me to get organized, and for editorial responses that were at once sympathetic, challenging and inspiring. Mark is the editor we all dream of.

While Mark pushed me to think afresh about gender, technology and history within the great sweep of Asia's transformations, my colleagues in the Society for the History of Technology encouraged me to argue the case for bringing Asian perspectives into the broader history of technology. Here, my thanks are due in particular to Daryl Hafter, John Staudenmaier, S.J., Pam Long, Bob Post, Nina Lerman and Suzanne Moon. I would also like to thank the National Science Foundation for funding that allowed me to present a keynote paper on 'Significant technologies' at the SHOT 50th Anniversary National Science Foundation Workshop, 'Thinking big in space and time', held in Washington DC in 2007, and Bruce Seely and the other members of the organizing committee for including me in the programme.

Dagmar Schäfer revived my spirits in hard times and restored my faith in the history of technology as a vital tool for exploring China's past. She and her colleagues at the Max Planck Institute for the History of Science offered stimulus, inspiration and rigorous critique as I reworked the chapters of this book. My thanks to the Institute for providing me not only with a research fellowship, but also with generous support for an international colloquium that Dagmar and I devised together on 'Technological cultures'.

My colleagues in Social Anthropology at the University of Edinburgh have been consistently tolerant of my unorthodox approaches to the discipline. I have learned so much, so enjoyably, from students and from staff over the course of seven years of arguments and discussions in classroom, seminar room and pub that I don't know where to begin my thanks.

At Routledge, Stephanie Rogers and Hannah Mack have been delightfully humane editors.

As ever, thanks to François for showing me where to look, to Ramon for being curious, and to Sandy, my lodestar.

PERMISSIONS

Cover: 'The ten labours of women', New Year print from Fengxiang, Shaanxi. Modern reprint of an early twentieth-century image, reproduced from James Flath's 'Nianhua Gallery' (www.history.uwo.ca/nianhua) with permission.

Chapter 1 is an abridged version of 'Technics and civilization in late imperial China: an essay in the cultural history of technology', originally published in *Osiris* 13 (1998): 11–33. By permission, Chicago University Press.

Chapter 2 is a greatly revised and augmented version of 'Instructive and nourishing landscapes: natural resources, people and the state in late imperial China', originally published in Greg Bankoff and Peter Boomgaard (eds), *The Wealth of Nature: How Natural Resources have Shaped Asian History, 1600–2000*, Palgrave Macmillan, 2007: 205–226. By permission, Palgrave Macmillan.

Chapter 3 reworks key themes from Part Two (Chapters 4–6), 'Women's work: weaving new patterns in the social fabric', of Francesca Bray, *Technology and Gender: Fabrics of Power in Late Imperial China*. © 1997 by the Regents of the University of California. Published by the University of California Press. By permission, University of California Press.

Chapter 4 is an edited version of 'Decorum and desire: an architectonics of domestic space in late imperial China', originally published in Hsiung Ping-chen and Lu Miaw-fen (eds), *Neo-Confucian Orthodoxy and Human Desires: Post/Modernity in Late Imperial Chinese Culture* (*Lijiao yu qingyu: qianjindai Zhongguo wenhua zhong di hou/xiandaixing*), Taipei: Zhongyang yanjiuyuan jindaishi yanjiusuo, 1999: 1–69. By permission, Institute of Modern History, Academia Sinica.

Chapter 5 is an edited version of 'Tales of fertility: reproductive narratives in Late Imperial China', originally published in Siegfried Zielinski and Eckhard Fürlus (eds), *Variontology 3: On Deep Time Relations of Arts, Sciences and Technologies in China and Elsewhere*, Kunstwissenschaftliche Bibliothek Volume 37, Cologne: Walter König Verlag, 2008: 93–114. By permission, Siegfried Zielinski.

Chapter 6 is an abridged version of 'Science, technique, technology: passages between matter and knowledge in imperial Chinese agriculture', originally published in the *British Journal for the History of Science* 41, 3 (September 2008): 319–344. By permission, British Society for the History of Science.

Chapter 8 is an abridged version of 'Agricultural illustrations: blueprint or icon?', originally published in Francesca Bray, Vera Dorofeeva-Lichtmann and Georges Métailié (eds) *Graphics and Text in the Production of Technical Knowledge in China*, Leiden: Brill, 2007: 526–574. By permission, Koninklijke Brill NV.

INTRODUCTION

The power of technology

With the proliferation of cheap printing and the spread of prosperity, by the late sixteenth century Chinese people at every level of society were conversant with a profusion of visual themes and images, religious and secular, diverting or didactic.[1] One that was familiar to everyone was a set of images of everyday technologies in action, depicting the tasks of growing rice and making cloth. One might encounter the vivid scenes carved on an elegant jade brush-stand or crudely brushed onto a piece of household pottery, in a fine album edition, a household encyclopedia or a cheap almanac. Emperors collected antique versions painted by distinguished artists, while peasant families pasted up rough but vivid woodblock renderings as New Year prints (Cover Illustration and Figure 0.1).

The original for these images was a pair of scrolls painted in the 1130s, just after the fall of the Northern Song dynasty in 1126, by an official named Lou Shou. Lou was a native of Jiangnan, the rich lands flanking the lower course of the Yangzi River, renowned for the sophistication of their rice farming and the quality of their silk. Lou had held posts all over China but now was back close to home, serving as the county magistrate of Yuqiang, not far from the new Southern Song capital of Lin'an (modern Hangzhou) in the heart of Jiangnan.[2]

Lou Shou's first scroll depicted the 21 essential technical steps in rice farming, showing ploughs, harrows and rollers, irrigation pumps and rice mills in operation. Most, though not all, of the workers in these scenes were male. The second scroll depicted the 24 steps of sericulture, including weighing and sorting silkworm eggs,

1 Clunas (1997); Brook (1998); Brokaw and Chow (2005).

2 The sources and other references for this section are all given in Chapter 8.

FIGURE 0.1 Irrigating the fields, *Gengzhi tu*, 1467 version.

picking mulberry leaves, feeding and cleaning the worms, and operating spinning wheels, reeling machines and looms; here, most of the workers shown were women. Each scene was accompanied by an evocative poem. Lou Shou named his masterpiece 'Pictures of tilling and weaving', *Gengzhi tu*. This may seem a straightforward enough description of the subject matter, yet the words held a deeper resonance – to which we shall return.

In around 1154, on the advice of an imperial inspector to whom he had shown his paintings, Lou Shou presented the *Gengzhi tu* to Emperor Gaozong, the first ruler of the new and still beleaguered Southern Song dynasty. The emperor was impressed, indeed delighted, with Lou's offering. According to Lou's nephew, he 'praised and rewarded Lou, and gave orders for the scenes to be copied on the screen of the Inner Court with Lou Shou's name on them' – a signal honour. Lou Shou's work soon became famous among the ruling elite as a perfect medium for "promoting agriculture", *quannong*, one of the fundamental priorities of the imperial state and its officials at every level and a matter of particular urgency in the decades after 1127, when the Song government had been evicted by its enemies from all its northern territories.

Through the Southern Song (1127–1279), Yuan (1279–1368) and Ming (1368–1644) dynasties, the *Gengzhi tu* was copied or reworked by court painters, circulated in woodblock form by local magistrates, and incorporated into the imperial encyclopedia *Yongle dadian*. The popularity of the *Gengzhi tu* was not restricted to state officials, however. With the flowering of popular print culture from about 1550 on, variations on its images and poems came to feature in works such as farming manuals, primers and household encylopedias. Craftsmen had adopted Lou Shou's striking depictions of familiar scenes into their repertory earlier still. Starting in the late Song, we find versions of *Gengzhi tu* images painted, carved or stamped onto porcelain vases, pottery pillows, lacquerware, New Year prints and fans. They inspired lively local variants in Korea and Japan. They even became familiar to Europeans during the seventeenth- and eighteenth-century vogue for *chinoiserie*, since they were favourite choices for decorating export wares such as porcelains, Coromandel lacquer screens and printed wallpapers. The zenith of the *Gengzhi tu*'s success came during the early Qing dynasty (1644–1911), when no fewer than three emperors, Kangxi (r. 1662–1722), Yongzheng (r. 1723–1735) and Qianlong (r. 1736–1796), collected and studied ancient versions, composed new sets of poems, and commissioned new painted versions or stone carvings, which they had printed in woodblock format by the imperial printing house and circulated to officials around the empire. What is striking about these new versions, however, is that although they were considered potent instruments for imperial policies to "promote agriculture", in technical terms they depicted none of the improvements that had taken place since Lou Shou's time.

What accounted for the immediate success of the *Gengzhi tu* and for its enduring popularity? What meanings did it convey to readers or viewers, high or low? It depicts what today we would call technologies, but in what sense was it a technological document? How did its efficacy, the work that it did, shift over time and space or in different social contexts? And what possibilities does the story of the *Gengzhi tu* suggest for rethinking technology as a historical heuristic? I open and close my introduction with the case of the *Gengzhi tu*, a twelfth-century official's delicate and closely observed delineations of rural technologies that over the centuries became common visual currency, because it neatly illustrates the three main arguments of this book.

The first is that technology matters in Chinese history, in fascinating and complex ways that we are only now beginning fully to explore. If we take the anthropological view that technologies exercise symbolic and social as well as material efficacy, then we see their significance in a new light. In particular, I suggest that late imperial sources offer rich insights into the interplay between the micro- and macro-politics of everyday life and the workings of governmentality through the medium of technology and its representations. Second, focusing in this way on what Foucault called "techniques of the self" and "technologies of the subject" also allows us to develop new insights into gender, including the dynamics of gender regimes as they translated into material activity, and to show how gender principles were woven into the very fabric of empire.

My third point concerns time, history and narrative, and the place of technology in the "divergence debates". All too often the history of technology has been marked by Eurocentric teleologies, as has the genre of comparative history or sociology that mobilizes technology to explain historical change.[3] Other societies are then measured against a narrative of progress as calculated retrospectively by modern Western standards. The model of change is linear: either a society progresses along the predestined line, or it diverges, or it stops still. A great deal of the history of technology (and of science) in China has been concerned with explaining why China did not follow the same path as Europe: when and how did China take the lead? When and why was it overtaken? In consequence, the very term "technology" is tainted in the eyes of many historians of late imperial China: not only are its associations with Western supremacism distasteful and misleading, but the kinds of questions it raises appear largely irrelevant or alien to their own concerns.

Technology more broadly construed, however, need not impose narratives of progress or failure: the efficacy of a particular technology or sociotechnical system may lie, for instance, in complex long-term stabilizing effects rather than in dramatic ruptures. These more nuanced approaches pave the way for reintegrating technology as a mode of action into the broader currents of historiography. If one of the core challenges of late imperial history is, as Timothy Brook suggests in an earlier volume in this series, to rethink the loci, mechanisms and dynamics of state-society relations,[4] then I would suggest that technology provides a powerful lens for examining the groundings and workings of imperial governmentality. The following chapters explore just a few of the many ways in which material settings and practices (what Foucault refers to as *dispositifs*) shaped topographies of everyday experience, contributing both to the constitution of subjectivities and to the weaving of relations of power.

What is it, then, that we are talking about when we discuss technology?

Technology is, as Leo Marx puts it, 'a hazardous concept', [5] a foxy term hedged about with modernist assumptions and agendas. From a historian's perspective it is not an actor's category – certainly there is no close equivalent to the term in pre-modern Chinese.[6] Yet I do not think we should therefore reject it as a heuristic tool.

3 See Staudenmaier (1985) on the grand narratives characteristic of the history of technology, and Goldstone (2000) on the teleologies of comparative history and their place in the "divergence" debates.

4 Brook (2005b).

5 Marx (2010).

6 There were terms (not always systematically applied) for technical knowledge, skills and methods, for craft work, and for the general category of mechanical devices or machinery (this is discussed further in Chapter 6). None of these, clearly, is equivalent to the modern concept of technology; the contemporary term *jishu* was only introduced via Japan in the twentieth century. We should remember, though, that before the later nineteenth century there was no equivalent in European languages either; instead people referred to crafts, mechanical arts, or *arts et métiers*.

From an anthropological perspective, all societies have technology, that is to say, they possess repertoires of skilled material practices and artefacts that serve to produce food, shelter and clothing, to communicate, to control, to worship, to fight or to trade. Furthermore, they will have characteristic ideas about the forces mobilized by technological activities of various kinds; about how, whether and to what ends they should or should not be used; the nature of their effects; and their significance – understandings that STS[7] scholars refer to conveniently as *technological cultures*, a close equivalent to Lewis Mumford's earlier concept of *technics*.[8] As George Basalla puts it, a human technology is a 'material manifestation of the various ways men and women throughout time have chosen to define and pursue existence'.[9] The materials and skills that make up any technological practice – cooking, say, or building bridges, or transporting people from home to work – are exercised within a social context, imparting meanings both to the objects produced and to the people who produce them. Technologies in this definition are specific to a society, embodiments of its visions of the world and of its struggles over social order. In this sense the most important work that technologies do is to produce people, and relationships between them. How did other societies see their world and the human place in it, what were their needs and desires, and how did the technologies they developed serve to fulfil them?

We can think of any technology, present or past, as a network of objects, activities, knowledge and meaning, comprising: goals to be met; tasks to be done; artefacts made for these purposes; the uses (material, social or symbolic) to which the artefacts are put; the resources and processes by which they are produced and made available; and the skills, perceptions, knowledge and organization mobilized in these processes – including ritual, magic, arts and social techniques as well as skills in the manipulation of physical materials, and indeed of the human body. The body, as Marcel Mauss observed, serves us both as a tool and as an object of technical action.[10] The acquisition and practice of technical skills trains our muscles and perceptions and helps to define who we think we are and how we see the world. As well as these incidental effects, we have deliberately devised innumerable techniques and tools to rework our bodies, to shape, maintain or transform them, ranging from deportment or clothing to weapons, surgery or reproductive technologies, ancient and modern. Whether from a Marxist, Durkheimian or

7 Read alternatively as 'Science, Technology and Society' or as 'Science and Technology Studies'.

8 See Mumford (1934) on technics, and Bijker (2007, 2010) on technological cultures. Although Bijker suggests that technological cultures are specific to modern societies in which technology is explicitly conceptualized as a powerful mode of human action, in my view it also constitutes an effective tool for investigating historical societies (e.g. Bray, 1999: 163).

9 Basalla (1988: 15). For a discussion of the distinctions between animal and human technologies, see Aunger (2010).

10 Mauss (1950).

Foucauldian perspective, in technology studies today the body as artefact and instrument is beginning to occupy a central place.[11]

At the same time, sociological theories of technology in modern industrial societies have recently converged with archaeological and anthropological analysis to treat making, using and doing as inseparable elements of technological practice. The old analytical chasm between invention and routine, the metaphorical hierarchies of head (engineer), hand (factory worker) and stomach (consumer), have given way in mainstream history and sociology of technology to investigations of the dialectics between making and using, with the understanding that skills, judgement and choice are involved at every step – and that bricolage, negotiation, creative compromise and distributed cognition are at the core of all technical activities, including the most sophisticated modern engineering projects.[12] This broader and more realistic definition of what counts as technological activity and where it is located challenges stereotypes of technology as primarily a domain of white male expertise and draws attention to the fact that technical change is by nature systemic.[13]

I have just suggested that all technical practices are embedded in, or constituted by, a network or constellation of materials, procedures and relationships – what many STS scholars refer to as a *sociotechnical system*, a "seamless web" in which 'it is never clear *a priori* and independent of context whether an issue should be treated as technical or social'.[14] The advantages of this perspective are that it draws

11 On the legacy of Marx's insights into technological discipline, reification, deskilling and subjectification, and his own related concept of "technical code", see Feenberg (1999, 2010). On the genealogies of the Maussian-Durkheimian tradition of cultural technology and its relation to Marxism, see Warnier (2009). Warnier (ibid.) and Naji and Douny (2009) both discuss the impact of the shift from the Durkheimian "person" to the Foucauldian "subject", and the incorporation of "techniques of the subject" into recent anthropology of technology.

12 "Distributed cognition" refers to the fact that technological problem-solving involves (often unacknowledged yet essential) dialogue between the various human actors, while also mobilizing forms of tacit knowledge embodied in tools, instruments or skills (Hutchins, 1995). An impressive application of these ideas to a historical case is Chandra Mukerji's recent study of the building of the Canal du Midi (Mukerji, 2009). Actor Network Theory (ANT) offers an alternative way of theorizing this insight into the collective production of expertise in emphasising the importance of "enrolling" both human and non-human actors. Although most ANT studies address contemporary projects, it has been applied historically to dramatic effect, as in John Law's analysis of the Portuguese carrack and its place in a 'revolution in the means of long distance control' that transformed the balance of power across the globe (Law, 1986).

13 I discuss the links between these new approaches and feminism in Bray (2007c). Lerman (2010) documents the steps that went into making technology a white male preserve in the industrializing USA.

14 Bijker (2010: 67). The concept of sociotechnical systems was first elaborated by Thomas P. Hughes (1983) in his study of Edison as inventor-entrepreneur and the contrasting processes of electrification in the USA and Germany. The symbolic and phenomenological dimensions of sociotechnical systems have been of particular interest to anthropologists of technology (e.g. Ingold, 2000; Pfaffenberger, 2001; Bray, 2007c).

attention to the full range of resources upon which a technical activity depends, highlights the social identities of and relations between the actors involved, and helps us to structure comparisons within or between societies. Another, complementary analytical approach (first developed by French anthropologists of technology and now attracting cross-disciplinary attention) is to map the resources and procedures – material, social and symbolic – that go into making an object or performing a technical task as a *chaîne opératoire* or operational sequence. This draws attention to the fact that technologies are not simply material transformations, but also involve what Jean-Pierre Warnier labels "ritual efficacy", including both the production of subjectivities and the mobilization of cosmic energies and principles (whether construed as the physical laws governing mechanics, for example, or as the natural patterns of *qi* transformation).[15]

Where do we draw the boundaries around a sociotechnical system, or locate the limits in time, space and scope of an operating sequence? An observer or researcher may decide to "cut the network" differently depending upon his or her resources and his or her unfolding understanding of what is at stake.[16] A long yam in the New Guinea Highlands may take just a few months to grow, but the gardening and ritual cycles in which it figures stretch over much longer periods. How does the ethnographer, who only has one year in the field, decide upon the time frame (or frames) of the study? What goes into yams beyond well-hoed soil and careful weeding? Only after some weeks does the ethnographer start to wonder if lighting a fire on arriving at the field is significant; only after a few months does he or she begin to understand how the yams benefit from the farmers' betel-nut breaks.[17] Was the birth of a son in a late imperial Chinese home simply the happy result of a fruitful sexual coupling? Far from it, we realize, once we start to think about the operating sequence that went into the production of heirs. As I show in Chapters 1 and 5, in order to ensure this fortunate outcome a range of technical expertise was set in motion long before the parents fell into bed together. The boy's great-grandfather's grave was carefully sited by a skilled geomancer; daily offerings were made at the ancestral altar; his parents' horoscopes were rigorously matched during the negotiation of the betrothal; the carpenter who built his mother's marital bed performed the correct rituals during its construction and minutely

15 This form of mapping begins with systematic observation of the operational sequences of production or use, and ideally proceeds to analysis of what Pierre Lemonnier (1992) calls the "social representation of technologies": this denotes not only the kinds of meaning that usually attract the attention of cultural anthropologists, but also the ideas governing the construction and use of tools and artefacts, an ethnoscience of material nature and action that includes their ritual dimensions and the ways in which material action forms subjectivities; see Pfaffenberger (2001); Mahias (2002); Bray (2007c); Coupaye (2009); Warnier (2009); Dobres (2010).

16 Strathern (1996).

17 Coupaye (2009).

calculated the measurements of the timbers; his mother took care to regulate her menses, without which medical theory deemed that conception was impossible; his father might well have taken medical theory on board too and cut back on the frequency of his sexual activities to improve his chances of impregnation.[18]

While it may not be possible to retrieve all the elements involved in a historical case of an operational sequence or sociotechnical system, such approaches remind us that we cannot simply restrict to material factors our consideration of the work that technologies do or the resources they require, nor will the same clusters of techniques, concerns or forces necessarily be at stake in what appear to be identical technological enterprises undertaken in different social settings.

At this point I may appear to be advocating an approach to technology, in tune with the current "cultural turn", that plays down or even denies technology's material dimensions and exigencies. We see a paradoxical denial of matter, for instance, in the recent obsession with "materiality" in material culture studies. By and large, scholars in this burgeoning field treat artefacts as ready-mades or black boxes, essentially discounting their material features or constraints – whether at the stage of production or in use – to focus on the semiotics of consumption or interpretation. In this analytical vein the artefacts themselves become empty vessels to be filled with meaning, almost infinitely malleable in their interpretation.[19] But I disagree fundamentally with this extreme interpretivism.

I certainly wish to stress that any material artefact can and usually does accommodate more than one register of meaning, which is one reason why our stuff can shape our lives and beliefs so effectively.[20] But I also insist that we cannot understand the choices people make about technology unless we pay due attention to the specific material characteristics that shape or constrain how we humans occupy, respond to and shape our environment.[21] Nuts and bolts must fit together; an engineer's choice between iron and steel is not arbitrary, and wool will not do as a substitute. A ceramic cooking pot can be round-bottomed or flat, coiled or made on the wheel, rough or glazed, a one-off or mass-produced, but if it is to function as a cooking pot, then the potter (or manufacturer) must fashion and fire suitable clay in such a way that the contents do not leak or drop out.[22] A society

18 See also Bray (1997: Chapters 7–9).

19 Daniel Miller, the leading theorist of material culture studies, explicitly dismisses technology as an analytical category, restricting his interest instead to how technologies such as the Internet or cellphones are brought into being as cultural artefacts; e.g. Miller (2005); Miller and Slater (2000). For critiques of this approach, see Pinney (2002); Bray (2007c); Ingold (2007).

20 I discuss the power of such ambiguities in almost all the following chapters, but most explicitly in Chapters 1 and 4 on domestic space.

21 See the essays on technological choice in Lemonnier (1993), which span the human experience from the neolithic to the nuclear age.

22 Mahias (1993); van der Leeuw (1993).

that does not possess the technical knowledge and equipment for high-temperature firing simply cannot produce steel or porcelain; furthermore, true porcelain depends for its translucency on the use of a very special kind of clay – no other clay will do, as numerous frustrated European sovereigns and technicians came to realize over almost three centuries of fruitless experiments.[23]

To understand the technological choices that a society (or an individual) makes about how to solve a problem, then, we must consider how both material and symbolic efficacy are merged within that choice. We also need to consider what we might call the logic of development of particular technical or sociotechnical systems, and the extent to which technological choices are "path dependent", inflected by earlier choices and the infrastructure and styles of technical solutions already in place. A society that has invested heavily in large-scale water control, for instance, or in automobiles or nuclear power, will not find it easy physically or economically to shift to another system, even when the disadvantages or dangers of the current system become blindingly clear.[24]

In the case of imperial China, the path-driven logics of small-scale rice farming or of large-scale hydraulic projects for flood control and irrigation cannot be ignored, yet (as I discuss in Chapter 2) they should not be treated as self-contained, as homogeneous through space and time, or as inflexible technological determinants that ruled out choice – as in Wittfogel's notorious equation of large-scale hydraulic projects with oriental despotism and unchanging Asiatic modes of production.[25] It is not only modern, scientific societies that are able to manipulate and inflect complex infrastructures and their needs. We can observe, for example, some powerful parallels in the choices made between alternative hydraulic cultures in Ming-era Jiangnan, and in the twentieth-century West. Exploring the technological cultures associated with hydraulic policy in the United States and the Netherlands post-World War II, Wiebe Bijker contrasts what he calls emergency and stability approaches to conceptualizing and solving large-scale hydraulic problems (no prizes for guessing which approach predominates in which nation!). Charting the interactions between state, provincial and local hydraulic expertise and policy in Ming Jiangnan, Cho-ying Li similarly identifies two contending models for dealing

23 See Wagner (1993, 2008) on iron and steel in China and on the connections between high-temperature-fired ceramics and the development of steel technology. See Liu (1999) and Finlay (2010) on the long European quest to recreate porcelain.

24 On path dependence or dependency, economic and technical, see Puffert (2008). Wagner (1997) documents an intriguing reversal of this logic, showing that the industrial-scale centralized system of iron production that predominated in early Chinese dynasties was progressively abandoned in late imperial times and replaced by small-scale local smithing.

25 Wittfogel (1957); see Brook (1999, 2005b: 7–10) on Wittfogel's place in the lineage of Western conceptions of oriental despotism, and on the impact of his interpretations on the historiography of imperial China.

with the threat of floods in the region: one model prioritized dredging watercourses; the other prioritized careful maintenance of the dykes. The first model was basically a response to emergency conditions; the second aimed to maintain long-term stability. The first was largely favoured by the central state because it required fiscal and administrative inputs only in times of impending crisis; the second, which helped ensure reliable harvests and reduced the risk of catastrophes, was naturally considered vastly superior by local farmers.[26]

More typically, however, the material worlds of other societies embody quite different values or goals from our own. I suggest that a focus on *significance* can be very useful to the historian of technology.[27] Technologies that we regard as crucial may be disregarded or hold different meanings elsewhere, while others that seem to us of negligible importance may hold great symbolic significance, as in the case of butter-making and fire-lighting in India, where the housewife daily re-enacts the creation of the cosmos under the family roof, or of the construction of the Chinese ancestral shrine.[28] Within a single location the same technology may change its meaning radically over time: in Classic Maya society weaving defined class, in Aztec Mexico it defined gender, and in contemporary Meso-America it defines ethnicity.[29] Conversely, the significance attributed to an activity may remain constant while its social and material organization is transformed – as in the equation of weaving and womanly work in China, which I discuss in Chapter 3.

From a historian's perspective the concept of significance can usefully be applied at two levels, which anthropologists would call *emic* and *etic*. We can try to locate technology's significance emically, as an expression of a society's own goals and values, and/or etically, as a characteristic that shaped that society's distinctive characteristics or its historical development. We may deduce that a technology was significant within a society if one or more elements in the network of goals, tasks, processes and artefacts just mentioned are highly visible in the historical record, whether as material remains; explicit subjects of description or debate; or taken-for-granted components of key activities, whose significance has to be read against the grain. Let me illustrate this point briefly with two examples that are not included in the chapters below.

First, reading against the grain. An American asked to name the most significant technology in their country today would probably fix on the computer industry or biotechnology, space capsules or nanotechnology. Yet from the viewpoint of

26 Bijker (2007); Li (2010).

27 It is the feminist scholar Autumn Stanley who has argued most explicitly for the importance of 'redefining technology' and 'redefining significance' as a way towards a less-biased understanding of the part technology has played not simply in Western but in human history (Stanley, 1995: xxxi).

28 See Mahias (2002) on the cosmic efficacy of women's work in India, and Chapter 1 on the significance of altar-centred architecture in China.

29 Brumfiel (2006).

an anthropologist the flush toilet immediately suggests itself as the technological object that best encapsulates the modern American way of being-in-the-world.[30] It keeps people clean and fragrant and spares them all direct confrontation with the unpleasant by-products of their consumption. Defecation is taken for granted as the most private of activities, yet individual privacy depends on a vast infrastructure of water and sewage networks maintained at huge but unacknowledged public cost. Males and females have separate public toilets with different hardware, different toilet practices – and (as quickly becomes apparent in seminar discussions) *very* different views of how toilets need to be cleaned. When Americans venture abroad, the perceived barbarity of foreign toilets (including German toilet bowls designed to allow inspection of the results, Japanese Toto toilets that draw attention to unmentionable body parts by washing and blow-drying them, or dual flush buttons that require the user to distinguish between different kinds of elimination) reaffirms a sense of civilizational superiority. American travel stores are full of devices and booklets for coping with foreign toilets or lack thereof, and one nationally syndicated radio series devoted 10 minutes each week to horror-filled anecdotes of "going abroad". A technological history of modern American society that took the bathroom as its key focus would not invalidate the interpretations that studies of electrification or the rocket industry suggest, yet in terms of emic significance it would tell us as much if not more about ordinary, everyday life and people's deep-rooted but seldom examined values.

My second example concerns bronze-making in ancient China, and the significance of bronze bells in particular.[31] Shang and Zhou China (c. 1650 to 256 BCE) produced vast quantities of bronze objects, almost exclusively for the use of the ruling class. Many magnificent ritual vessels and weapons were buried with their owners and have survived to the present. Bronze bells were among the most beautiful and elaborate artefacts retrieved from the tombs, sometimes in sets numbering several dozens. The material resources and technical sophistication that went into designing and firing the complex, multiple-component, delicately engraved ceramic moulds, into the casting and assembly of the huge pieces, and into their accurate tuning (the most sophisticated bells were designed to ring two distinct notes) were prodigious. Huge workforces organized into small, specialized teams were involved in what Lothar Ledderose describes as "making bronze the complicated way".

In terms of emic significance, within the worldview of the time, Shang and Zhou bronzes were significant as both badges and instruments of rule. Paramount rulers bestowed bronze vessels on lesser feudal lords along with allocations of territory

30 Bray (2008a). The essays in Dutton *et al.* (2002) explore some non-American toilet cultures, past and present, and the ways in which they construct civilizational hierarchies.

31 This discussion is based on the work of Lothar von Falkenhausen (1993) and Lothar Ledderose (2000).

or subjects; fine bronzes were exchanged between the aristocratic class, cast to celebrate conquests, and passed down as heirlooms. The inscriptions on the bronzes, and other surviving texts of the period, say almost nothing that relates to how they were produced, although their production involved the organization of vast resources knitting together ruler and subjects through a closely controlled system of production that enforced the social hierarchies of a feudal world. For the Shang and Zhou ruling class the bells were especially significant as crystallization of political, social and cosmic power: played on the occasion of ritual feasts, they communicated with the ancestral spirits to call down their blessings upon the living. Their notes were unchanging, symbolizing constancy and transmitting a cosmic order that resonated through the human world:

> With rites and music, [the Minister of Rites] adjusts the transformations of Heaven and Earth and the production of all the material things, so as to serve the ghosts and spirits, to harmonize the multitudinous people, and to bring all the material things to perfection.[32]

In ancient China, as in ancient Greece, music played in the wrong tones could cause cosmic and social disruption. Unlike stringed instruments, bells do not go out of tune, so they set the standard within an orchestra. Tuning each individual bell to maintain consistency within and between sets of bells was thus of the utmost cosmic and political importance. According to the philosophy of the high Zhou period, between about 1000 and 500 BCE, only those of high moral worth as well as social rank were fitted to play musical bells, and the cosmic power of the music produced by the bells warranted devoting huge resources to their design and production. But by the time of Confucius, around 500 BCE, many people were starting to criticize such huge investments in bell-sets as an unnecessary luxury; rather than an essential practice of rule, listening to bell music began to be thought of as a sensuous self-indulgence that distracted rulers from their proper duties. By the time that China was united as an empire in 221 BCE, bronze bells had been discarded from orchestras to be replaced by stringed and wind instruments.

What, then, was the longer-term significance of Shang and Zhou bells, and of the wider practice of what Ledderose calls "modular production" of bronzes? The bells were among the consummate products of the bronze-maker's art, epitomizing the complexity and sophistication of craft skills of the era. Initially developed to serve the ritual needs of a feudal society, this Chinese system of modular production – whereby models or templates were disaggregated into multiple components,

32 *Zhou li* (Ritual system of the Zhou), "Minister of Rites", quoted von Falkenhausen (1993: 2).

allowing for relatively easy series production and for the recombination of a few standard elements to produce apparently spontaneous and unique variations – proved immensely durable and adaptable. To cite just two of Ledderose's sequence of examples, in the third century BCE it was used to produce the famous Terracotta Army for the First Emperor's tomb, and in the Ming and Qing, almost two millennia later, it was applied to the mass manufacture of fine porcelains. A technique initially rooted in ancient modes of exploitation and consumption evolved into a tool of imperial rule and a catalyst of commerce, shaping Chinese patterns of production – and ways of thinking about how to design and produce in mass, and how to exercise quality control throughout the production process – right through to the modern era.[33]

In my own search for significant technologies in late imperial China I have usually looked for categories of material production and practice that have come down to us in documentary form as prominent concerns in debates of the time. This of course biases me towards the concerns of the literate male elite – but it gives me somewhere to start, and confronts me immediately with interpretive issues related to governmentality: how to locate normative or prescriptive texts in a landscape of practice, for instance, or what sense to make of contrasting politics of representation in different genres of writing about a given domain of technology. By this reckoning, several of the technical domains that we now consider fundamental (such as metallurgy or engineering) were at best marginal in late imperial debates.[34] Others, such as agriculture and textile production, were central concerns, yet as I discuss in Chapters 2, 3 and 6 their local meanings and value cannot be adequately understood in the terms of modern engineering and economics. Even texts that describe technological artefacts or practices in detail, many of which have been treated by modern scholars as technological texts – that is, as documents designed 'to transmit technological information to someone who [can] use it in a technological way'[35] – may on closer inspection turn out to have been very different in intent.

33 On the complex history of quality control methods in China and their importance as what Foucault (1973) would call "technologies of the subject", apart from Ledderose (2000: *passim*) see, for instance, Barbieri-Low (2007: 74 ff); Schäfer (2011b); Wang Guangyao (2010).

34 See, for example, the splendid critical review by Tim Wright on what can and cannot be pieced together from late imperial sources about iron and coal (Wright, 2007; also Wagner, 2008). Of course my approach also suffers from another bias, namely the privileging of written sources over material remains, with all the dangers this runs of taking intention or ideals for actual practice. Thus, for instance, the detailed rules laid down in the Ming Code for inscribing weaponry, boats, porcelain and other goods produced for the Court or for official use, were seldom followed in practice. Nevertheless, the symbolic importance of the regulations as a dimension of rulership was considerable (Schäfer, 2011b).

35 Long (2007: 6).

This is the case, for example, of the famous treatise *Tiangong kaiwu* (usually translated into English as 'The exploitation of the works of nature'), published in 1637 by a minor official of the late Ming named Song Yingxing.[36] The book contains detailed descriptions of the processes involved in a whole range of crafts and technologies, from growing cereals and weaving to salt extraction and sugar refining, the construction of boats, casting iron anchors and copper coinage, mining and pearl-fishing. Twentieth-century historians of science and technology in China, anxious to find roots of a modern mindset in the Chinese past, eagerly seized upon this work as testimony to a keen interest in technology among late-Ming intellectuals. For almost all of us engaged on Joseph Needham's great project *Science and Civilisation in China*, *Tiangong kaiwu* was one of the first texts we turned to.[37] With the turn to consumption studies in the 1990s, Ming specialists such as Craig Clunas challenged the view that texts such as *Tiangong kaiwu* were "how-to" texts, suggesting that they were aids to the appreciation of artefacts rather than blueprints for producing them.[38] Dagmar Schäfer's recent study supersedes the consumption-centred interpretations, which trivialize Song Yingxing's concern with the nature of technical processes. Locating *Tiangong kaiwu* in the 'scientific and technological landscapes' of the troubled years just before the fall of the Ming dynasty, Schäfer teases out Song Yingxing's accounts of craft techniques as part of his broader political enquiry into universal principles and specific modalities of cosmic transformation ('the works of Heaven', *tiangong*), and the material, social and moral consequences of different forms of human intervention ('the inception of things', *kaiwu*). In modern terms, then, we might be tempted to categorize *Tiangong kaiwu* as a scientific investigation of natural forces, explored through the energetic transformations manifest in different craft technologies. Where this interpretation falls short, however, is that for any Ming reader the linkages between microcosm and macrocosm, physics and ethics, an artisan's work and the political order, were self-evident, even if few readers of the time were apparently enthralled by the novel and challenging ways in which Song Yingxing linked cause and consequence.[39]

My own research reveals the epistemological importance of analogous moral and cosmological concerns in late imperial writings on "agriculture", *nong* – but shows that the significance of the activity's cosmic and moral efficacy was unequally

36 See Sun and Sun (1966) for an English translation of *Tiangong kaiwu*. Pan Jixing is the outstanding expert on the work; see Pan (1989) for a comprehensive historical account and textual analysis.

37 My own contribution to the Needham project was the volume on agriculture (Bray, 1984).

38 Clunas (1997).

39 Schäfer (2011a). Analogies – but also stimulating contrasts – can be drawn with Europe at around the same time, including the productive synergies between craft and natural philosophy, as well as the non-materialist dimensions of both activities (e.g. Roberts, 2007).

weighted in different genres of technical writing. While these concerns were very prominent in the technical accounts produced by or for officials, they were – as I discuss in Chapters 6 and 7 – at best muted in the numerous farming treatises written by private landowners, which were highly pragmatic and business-like. In the eyes of the governing elite, agriculture was the "fundamental occupation", *benye*. Of course its material efficacy was of primary importance to officials: it produced the basic requirements for human subsistence and for the payment of taxes (grain and cloth), as well as all kinds of other products that could be consumed by farmers or circulated through the market. As such it was treated in official texts as a sector that could be organized more or less productively in technical, economic and environmental terms. But at a higher level the proper organization of farming was viewed as the basis of the ideal relationship between ruler and subject, the root of the state's strength and stability, and the confirmation of a moral order in which selfish striving after profit was rejected in favour of modest and stable prosperity. Equally important were the cosmic resonances of the proper divisions of labour, between classes[40] and perhaps most notably between the sexes: as the classic saying went 'men plough and women weave', *nangeng nüzhi*: working harmoniously together, farming couples fed and clothed their own families and supplied the empire, while the complementarities of their humble everyday work reproduced the fruitful conjunction of *yin* and *yang* energies that underpinned the political and cosmic order. Much of the knowledge produced and circulated about agriculture in late imperial China fits neatly into modern explanations of materialist efficacy – indeed that is how I myself treated it when I was beginning my research. Yet, as I came to realize, only if we bear in mind the cosmo-political significance of agriculture, in particular its significance as a gendered activity, can we satisfactorily account for certain registers of representation or policy choices in the Chinese case. (Agriculture remains equally charged today, politically, emotionally and symbolically, as witness disputes over the European Union's Common Agricultural Policy, or the struggles for and against genetically modified crops.)

Much of my recent work has explored how we might use technology to develop new insights into gender, including the dynamics of gender regimes and the impact of gender principles as they translate into material activity. The field of feminist technology studies has transformed our understandings of the place and power of technology in the modern world, but there is no reason to restrict the scope of such analysis to the industrial age.[41] In late imperial China, as the following essays

40 Maintaining the correct divisions between the four classes of occupation, *simin* (scholar-official, farmer-peasant, craftsmen and merchant), was a recurrent topos in late imperial discussions, especially at times when commerce seemed to be encroaching rapidly into people's lives (e.g. Brook, 1998; and see my discussion of the rubric of *li* (benefit or profit) in Chapters 6 and 7).

41 Lerman *et al.* (2003); Bray (2007c).

show, we can see gender principles at work in such diverse fields of technical activity as the forms and uses of domestic architecture; changes in the organization of textile production and the perceived material and moral stakes involved; or contrasting styles for the inscription and dissemination of farming skills. Whether through the construction of material environments, the production of economic or social goods, or the values and efficacy attributed to different sorts of technical skill and knowledge, technology played a key part in shaping and defining the nature, range and interrelations of masculinities and femininities characteristic of late imperial China, and their evolution over time. Technology wove gender principles into the very fabric of empire in China, from cosmology and ideologies of rule to the material foundations of the state and the everyday practices of the domestic sphere.

Technology and gender are always mutually constitutive, and thus for the historian as for the sociologist they should, sources permitting, be mutually illuminating.[42] But there is still a widespread tendency even within the academy to presume that gender means women. There are two problems here. First, technology has usually been thought of as a masculine domain of competence, and it is only recently that feminist historians have succeeded in showing the value of treating women as key participants in the history of technology.[43] Second, the relation between men and technology has usually been taken for granted as quite obvious, and it is only quite recently that scholars of technology have started to inquire critically into how particular technological cultures shape corresponding forms or variants of masculine identity.[44] In arguing for bringing together gender and technology in the case of late imperial China, I am not just arguing that technology has a place in developing "women's history" by reconfiguring the analysis of female subordination and agency, although that is certainly true. Equally important, in exploring the material constituents of gendered subjectivities and action, a technological perspective raises questions – questions that might not be obvious from the perspective of, say, legal or social history – about the nature and range of masculinities as well as femininities within a society. A focus on technology can cast light not only on the material workings and attributions of male-female difference, but also on male-male and female-female difference, and on tensions and shifts in gender regimes over time. It prompts us to consider the material impact of different forms of gendered action, and how this translates into real or perceived contributions towards building, maintaining or challenging the social order.

42 'Since technology and gender are both socially constructed and socially pervasive, we can never fully understand one without also understanding the other' (Lohan and Faulkner, 2004: 319).

43 And of tracing the processes through which technology came to be identified with (white) men and with machines (Adas, 1989; Lerman, 2010).

44 Horowitz (2001); Mellström (2002); Lohan and Faulkner (2004).

As in any society, in China certain technical skills and responsibilities, certain domains of material experience, were associated with men, others with women, and some with both sexes. Class, rank and age complicated the schema, as did geographical or ethnic variation, and circumstance or the passage of time often brought actual practice into conflict with long-standing ideals. The significance of such attributions and conflicts was enhanced by a political philosophy that explicitly rooted cosmic order and the governance of the state in the ethical and material habits of the home. Although Chinese moralists and officials were unusually explicit and eloquent on these linkages, the idea that the overarching social order depends on the proper ranking of different kinds of work, or on the preservation of a particular domestic way of life and its associated technologies, is hardly unique to pre-modern China.[45] Yet the clusters of technological practice that constitute what I call the *gynotechnics* and *androtechnics* of a particular society – sets of technological attributes and practices that combine to shape and define female and male roles and hierarchies – are much less obviously universal than one might at first think. Although they are in some degree specific to each society, however, they are by no means uncontested, uniform, or static. Their manifestations, range, interplay and change over time in late imperial China are pervasive themes in the essays that follow.

Let me mention immediately, however, one helpful clarification of late imperial gender regimes that a focus on technology provides. In common understanding, imperial China is notorious for what is supposed to have been its uniquely ruthless subjugation of women, conveniently symbolized by the lotus slipper. Late nineteenth-century critics, Western and Chinese alike, held that one reason for China's weakness and backwardness compared to the West was that half its population, namely women, were unable to contribute anything to society. Reduced by foot-binding and seclusion to helpless dependency, illiterate, silenced, lacking any moral authority, unable to engage in paid work or to provide any recognizably useful contribution to the family budget beyond subsistence activities, the poor creatures were valued at best for the sons they bore. Women crippled and enslaved could contribute nothing to national progress, and the reformers of the Republican period promoted unbound feet as an essential element of China's transformation into a modern nation, as did their communist successors. Meanwhile the stereotype of unmatched female oppression gained general currency at home and abroad, influencing professional as well as popular understandings of imperial society.[46]

45 Bray (2008a).

46 Ko (2005); and see Patricia Ebrey (2003: 194–219) for a useful survey of evolving Western views on how the Chinese treated women more generally, and what they felt this signified in comparison to their own customs.

Feminist historians have recently challenged the "May 4th view" of women in imperial society on a range of fronts. The point of this revisionist project is not to deny that late imperial society was organized on patriarchal principles, or that subordination to men shaped the lives of Chinese women and restricted their agency in most circumstances. (This is true, of course, of most women in all times and places, even in our own enlightened modern West.) Rather, the point is to challenge Eurocentric distortions and stereotypes, and to develop a more accurate and nuanced view of how patriarchy and gender worked in specific situations in China – taking into account the interplay with other classificatory categories such as class, rank, age or occupation, not to mention the various hierarchies and forms of subordination to which Chinese men were subject. Among the most influential revisionist studies are those documenting the moral authority that men attributed to women, their role as educators and managers, or the range of profitable work in which they might engage.[47]

In my own contributions to this feminist project, I set technology to work to develop some novel perspectives for assessing how the importance or unimportance of particular women was construed in late imperial society. I identified the nexus of technologies of space, work and reproduction (both biological and social) as a significant set, a *gynotechnics*, forming the material matrices within which virtuous and capable femininities were developed, expressed and judged during the late imperial period.[48] In unfolding how the relative importance of space, work or reproduction as the key locus of womanly worth shifted over time, between about 1000 and 1700, it became clear that patriarchal repression was just one strand in a much more complicated and interesting story. It is true that over time more Chinese women, across the social classes, came to live in increasingly strict spatial seclusion – an arrangement that today is often equated with social impotence. But in late imperial terms seclusion by no means signified that women lived cut off from the wider world. On the contrary, within the inner quarters women were responsible for producing a spectrum of social, ethical, ritual and economic goods, all of which were considered absolutely vital not only to the family budget and the general prosperity of society, but to the maintenance of the state and the cosmic order. In local terms, a competent and virtuous woman, even of humble family, sent out into the world valuable goods that included yarn or cloth or pickles or wine, and valuable persons, her well-brought-up daughters and sons. The ritual activities that she performed daily within the spaces of the house contributed, like

47 Ebrey (1993); Ko (1993); Bray (1997); Mann (1997); Pomeranz (2003). There is also now an extensive and entrancing literature on women as artists and writers, particularly interesting for the light that it throws on women's sociality, and on the various opportunities they had to leave their homes for visits or long-distance travel; see, for example, Fong (2008).

48 Bray (1997).

the successful raising of sons, to the well-being of the ancestors and the continuation and prosperity of the family line; and her strict observation of spatial boundaries further contributed to the family's respectability. From a statecraft perspective, meanwhile, women's material performance was considered an essential element of good government – not just complementary, but equal in importance to the work of men – and this was widely reflected in policy.[49]

From a modern economic perspective, as indeed from the emic down-to-earth standpoint of the family budget, women often made, and were recognized as making, significant contributions to household incomes and to meeting fiscal or other financial obligations. Until taxes in kind were abolished in the course of the sixteenth century, the tax burden on the men and the women within a household was equally apportioned. Men paid their dues in pecks of grain, women in bolts of silk, units that were notionally of equal value. But once taxes in kind were replaced by cash payments, the situation changed. Depending upon what kinds of commodities were produced locally, women might contribute only to their family's subsistence, or they might produce valuable goods for sale, or earn wages in a range of occupations. Spinning, weaving and sewing, as canonical female work performed within the home, were always approved as legitimate ways for women to earn money, but we also find endorsement, even praise, for less secluded female occupations, including regional industries such as tea-picking that took teams of girls and women far up in the mountains.[50] Indeed, there is some evidence to suggest that women's earnings in late eighteenth- and nineteenth-century China were closer to men's than they were in England – a point that Pomeranz uses in developing his interesting challenge to conventional interpretations of the "Great Divergence".[51] In sum, looking at the changing technological landscapes of late imperial times serves as an effective device not only for integrating women into the making of Chinese history, but also for rethinking what that history was. It would be difficult, however, to define this history as linear, or at any rate as following the lines that came to constitute normality in the West.

From this follows my third point, which concerns time, history and narrative – and how we might develop fruitful new approaches to using technology as a historical heuristic. The perspectives on technology that I propose engage with history, and with historical narratives, in ways that most historians of China would not immediately associate with the history of technology.

Technology as we think of it today, both as a way of acting and as a way of understanding, is at once the midwife and child of the modern industrial world.

49 See Chapters 2 and 3 on official attitudes to textile production.

50 Lu (2004) offers a finely nuanced reading of how women's work as tea-pickers was valued and represented.

51 See especially Pomeranz (2003).

As such it occupies a special place in modern ways of thinking, and a very special place in modern ways of constructing history. I suggested earlier that all societies possess technological repertories that reflect their goals and values, their perceptions of their material needs and their understanding of how such needs can best be met. What is distinctive about technological repertories in the modern world is their continual, ever-accelerating pace of change and penetration, along with a widely shared view that such change has positive social effects, and general agreement that substantial resources should be allocated to promoting and enhancing further change. What is furthermore distinctive is that, although there is no consensus on exactly what technology *is*, we do agree that it is a thing in itself, and as such a force to be reckoned with.[52] This understanding of the potency of technology, whether for good or for ill, is part and parcel of modern life, its technological culture. All around us messages abound that 'innovation in the mechanical arts is a – perhaps *the* – driving force of human history'.[53] In ordinary conversation as in political rhetoric the term "technology" serves as a keyword that conjures up dynamism and progress: imaginative inventions; engineering skills translated into sophisticated new machinery; artefacts and activities that underpin the sustained development of modern societies by producing ever more effective solutions to material problems, generating wealth, convenience and comfort in the process. Technology's achievements all seem to be about creating new powers and possibilities, solving age-old problems and opening up new worlds, catalyzing breakthroughs and radical transformations of the social and economic order that we happily refer to as "revolutions" (the first and second "industrial revolutions", the "digital revolution").[54]

Now, I completely agree with the basic premise here, namely that radical changes in technology are disruptive. More or different resources may be needed. New actors may be brought into the system. Accustomed actors may be brutally or subtly displaced, new spaces opened up, new skills acquired, new identities formed. Changes in technology can transform material worlds, create new visions, disrupt society for better or for worse – and it is these radical transformations that have conventionally attracted the attention of historians and social scientists, along with the conditions and capacities likely to generate transformative innovation. And yet

52 On the spectrum of views of the nature and scope of technology, across and within disciplines, see, for instance, Sigaut (1985), and the special issue of the *Cambridge Journal of Economics* that brings together views from philosophy, economics, anthropology, archeology, sociology and organizational studies (Faulkner *et al.*, 2010).

53 Marx (2010: 561).

54 While most people tend to think of technology as either morally neutral or a force for progress, more nuanced or indeed thoroughly negative views of technological dynamism also abound (Heidegger being one of the modern giants in the field of technopessimism).

at the same time technology can be a powerful force for cultural stability, since it creates material forms that embody shared values and beliefs, tying people into orthodoxy through their everyday practices.

One important reason why people often resist technological transformation is because it threatens what they believe is the right and moral way to do things; technology can be thought of as *concrete ethics*, a material habitus that naturalizes social relations and moral principles; I explore this idea further in Section I.[55] As Bruno Latour puts it, 'technology is society made durable',[56] and social systems, like eco-systems, are often highly resilient to disruptive pressures, including those exerted by technological change. Sometimes the disruptive forces unleashed by transformations in one technical domain will trigger transformations throughout a society; in other cases, however, the tensions and pressures built up in one technological domain may be absorbed or counterbalanced by developments in another, as for instance in the cases I describe below of the interplay between textile production and domestic architecture in China (Chapters 1 and 3).[57] This adaptive or conservative potential is an aspect of technology that has been much less carefully studied, although STS research on the dynamics of "path-driven" evolution or cases of "failed innovation" acknowledges the issues to some extent.[58] But technology is not interesting or successful only when it produces social or epistemological ruptures. No less energy and creativity go into maintaining continuity, stability and cohesion than into triggering disintegration or revolution, and no less careful explanation is required. [59]

As a discipline, the history of technology emerged as an enterprise that charted the roots and the rise of the modern West and its path to world supremacy.[60] The trajectory of early-modern Europe, ushering in a new era of rapid growth, competition and conquest, as well as radical and extraordinarily efficacious new attitudes towards "useful knowledge", has been treated by some historians as a

55 I take the term "concrete ethics" from a recent conversation with Suzanne Moon, but see also Latour (1992) and Bray (2013) on technical artefacts and moral delegation; Feenberg (1999) and Bray (2013) on technological codes; and Moon (2007) on ideologies of technological policy.

56 Latour (1991).

57 The theme is more fully developed in Bray (1997).

58 See, for instance, Pinch and Bijker (1987); Akrich (1993).

59 De Laet and Mol (2000) propose the concept of "fluid technology" to address some of these reproductive dynamics and the astonishing material and interpretive flexibility of certain highly successful artefacts. Emphasizing just how many "old" technologies are still in use today, often in creative combinations with more recent innovations, Edgerton proposes the linguistic concept of "creolization" as a useful way to explain how people's technological repertories and technological cultures are patched together from indigenous and external components over time (Edgerton, 2007a, 2007b).

60 Staudenmaier (1985); Pfaffenberger (1992).

miracle, by others as cultural predestination.[61] Divergences from this dynamic path of historical progress were usually characterized as blockage, lack or failure. No longer considered a useful question by most leading historians of Western technology, it remains a key preoccupation in comparative history or in popular accounts. How, when and why did late imperial China, or the Islamic world, diverge from the path of Europe? What went wrong? What miracle ingredient or happy contingency, intellectual, institutional or material, did they lack that led them to languish and stagnate rather than building on earlier achievements to break through into industrial triumph? The sad paradox of Joseph Needham's magnificent project *Science and Civilisation in China* is that its most famous legacy remains the "Needham Question": having argued forcefully that until about 1400 China led the world in scientific and technological creativity, Needham was faced with the puzzle of why China then fell behind, to be overtaken by Europe in science, in economic growth and in military and political power. Why did a scientific or industrial revolution not take place in China? This question continued to obsess Needham until his death, and still exercises a grip on the public imagination.[62]

As Nathan Sivin tartly noted some time ago, however, asking why something did not happen is far less likely to produce interesting history than asking what did – the principle on which most historians of China sensibly enough work.[63] Although they too must engage with modern concepts and categories such as gender, religion, art, philosophy, law, property, society or the state, many of which have also been mobilized in the past to bolster evolutionary arguments about Western superiority, today none of these terms seems quite so distorting of Chinese realities, so tainted with positivist agendas and freighted with negative baggage, as technology. In consequence very few historians of China – at least until very recently

61 See Jones (1981) for the foundational formulation of the "miracle" argument; David Landes is one of the most influential proponents of the cultural predestination hypothesis (classically stated in Landes, 1969). Joel Mokyr (1990, 2009) traces the emergence of "useful knowledge" as a powerful new concept, goal and method in Enlightenment Europe, and in particular in England. It is there that conscious efforts to fuse and develop the resources of natural knowledge, mechanical crafts and entrepreneurship had their most spectacular initial success, paving the way for what one might call a modern technocratic philosophy of governance. Goldstone (2000) provides a concise, if highly critical, overview.

62 Needham's culprits included the stifling effects of "bureaucratic feudalism" and a supposed indifference on the part of the Confucian elite to nature and practical knowledge (Needham 1969: 16, 190). The enduring power of the "Needham Question" as a popular plot-line is skilfully harnessed by Simon Winchester in his recent biography of Needham (Winchester, 2008, reviewed in Bray, 2010), and in more scholarly vein "the Needham Question" figures as the key theme of a recent edited volume on *Chinese Technological History* (Liu and Deng, 2009; O'Brien, 2009), and as an entry in the Chinese Studies section of the Oxford Bibliography Online that is currently in preparation.

63 Sivin (1982).

– have eagerly engaged with innovative or critical approaches within the history and sociology of technology as they have with new historiographical and interpretive approaches to the state, or money, or gender.[64]

Given how technology has been used and abused in constructing grand narratives for China's past, it is not surprising that many historians of China nowadays regard the term with misgiving, or dismiss it as the necessary yet culturally uninteresting nuts and bolts that hold a material culture or socio-economic system in place. As an example, in his recent study of artisans in early imperial China, Anthony Barbieri-Low – who identifies himself as a 'material-culture historian' – is at pains to draw a clear line separating technology from culture.[65] The category of technology he restricts to the material procedures and constraints of metallurgy or brick-making. Yet in unfolding how this domain of materiality, skill and control shaped the social conditions of production, the lives and identities of the workers and the significance and value of the artefacts produced, Barbieri-Low draws a magnificent portrait of what most historians of technology in the modern West would call a sociotechnical system.

There is one field of Chinese history, however, that still deals very explicitly with technology, and that attracts more attention outside the field of sinology than any other branch of Chinese history. This is the field of studies of economy, state and society in late imperial China that debates the place of China in the emerging capitalist world-system, between about 1500 and 1850. Here, late imperial technologies and technological cultures are a central focus of analysis, whether as vectors or inhibitors of growth or progress. The debate is framed overall by an elegant twist on Needham's question, set out by Mark Elvin in his hugely influential work *The Pattern of the Chinese Past*.

According to both Needham and Elvin, the heyday of Chinese scientific and technological achievement, and of its economic and intellectual dynamism, was the Song dynasty (960–1279).[66] Whereas Needham sought his explanations for China's perceived post-Song slowdown or decline primarily in intellectual predispositions, in *The Pattern of the Chinese Past* Elvin located the primary cause in what he argued were the *involutionary* dynamics of the Chinese mode of production.[67] Agriculture and the manufacture of commodities fell into what Elvin called a "high-level equilibrium trap": output could be almost indefinitely expanded

64 See Kuhn (2006). One inspiring exception is the oeuvre of Donald Wagner on metal, society and culture (e.g. Wagner, 1993).

65 Barbieri-Low (2007: 17).

66 Some historians cast doubt on whether the levels of economic and industrial productivity claimed for the Song dynasty were in fact quite as high or as novel as claimed by scholars such as Robert Hartwell and the historians of the Kyoto School, upon whose research Elvin and others based the idea of a Song economic revolution (Wright, 2007: 400–402).

67 Elvin (1973); the concept of involution is borrowed from Geertz (1963).

by increasing labour inputs, thus accommodating China's growth in population – at the expense, however, of diminishing returns. The kind of radical technological innovation that raises the productivity of labour and other inputs by a quantum leap – what Joel Mokyr calls the "free lunch" effect – never took place in post-Song China. One reason, Elvin has argued, was that the abundance of cheap labour removed any stimulus for major investments of capital in technological or organizational innovation. Another was the absence of any equivalent of early-modern European networks of science, and of the corresponding feedback between scientific and technological advance.[68]

Elvin's paradigm-setting study drew up a line of battle that continues to polarize historians today, provoking much serious rethinking of histories East and West. On one side of the debate, the historians of the "California School" argue that until at least 1800 the Chinese manufacturing, market and export economy was far from stagnant but grew steadily, remaining the largest and most productive in the world, dominating and shaping global exchanges and emerging institutions in ways that significantly challenge simple Eurocentric interpretations of the modern world order.[69] They nevertheless acknowledge that from about 1800 we see a "Great Divergence" between China and Europe (or rather Britain, the spearhead in the development of industrial capitalism). So what factors triggered the divergence? Had China by 1800 reached the limits of growth and entered a period of involution and diminishing returns, or was it the ruthless depredations of Western powers that reduced it to the status of "sick man of Asia"? Was Britain's rapid rise to world domination a happy accident, the product of coal and colonies, or was it firmly rooted in effective new attitudes towards "useful knowledge" that underpinned a radical break with earlier modes of production and accumulation?[70]

The California School believe that the rise of the West was contingent, not predestined, and most hold that under somewhat different circumstances China could have continued to prosper, although the trajectories of growth and accumulation that it had established would have led in different directions of development from those of industrializing Europe. They suggest that various aspects of lifestyle associated with the birth of the modern subject – widespread markets for consumer goods or managed birthrates, for instance – flourished in late imperial society even before they became established in Western Europe. Some members of the California School even propose that had China managed to maintain its dominance,

68 For a recent summary of his position, see Elvin (2008), in particular pp. 103–104 on "scientific thinking".

69 Landmark studies of the California School include R. Bin Wong (1997); André Gunder Frank (1998); Li Bozhong (1998a, 1998b); James Z. Lee and Feng Wang (1999); Jack A. Goldstone (2000); Kenneth L. Pomeranz (2000); and Giovanni Arrighi *et al.* (2003).

70 See especially Mokyr (2009: 40–62), 'Useful knowledge and technology'.

a kinder world might have resulted. Others note the resurgence of key features of late imperial social and economic organization that have recently helped propel China to a new position of dominance on the world stage.[71] Their opponents argue that late-Qing technology, institutions and intellectual practices had all outlived their usefulness, or insist that no society which lacked the systematic approach to advancing technical and scientific knowledge that had crystallized in Europe by 1800 could successfully have diverted its inevitable rise.[72]

Whether they accept or challenge the superiority of European civilization and its paths to dominance, almost invariably these studies assume that technology should be assessed essentially as a potential catalyst of change, whether in the economic sense of a production function, or more broadly as an intellectual project essential to the modern mindset of progress. Furthermore, even the scholars of the California School generally use the values and institutions of today's world to assess how close Qing China might have been to becoming modern under its own steam. Does this mean that the approaches that I have been advocating here rule out engagement with the Great Divergence debates? Let me first be clear that while I sympathize strongly with many of the points made and positions taken by the California School, I do not believe that but for coal or colonies imperial China might have beaten England to an Industrial Revolution. But like the California School historians I am deeply interested in how it was that China managed so much for so long.[73] And I share their interest in how technological choices and cultures shaped China's historical trajectories and its eventual encounters with the Western powers,[74] although (as I have already suggested) my view of what constitutes relevant technology for the purposes of such analysis, and how we should judge the goals and efficacy of the sociotechnical systems at issue, is by and large much broader.

Among the California School, Kenneth Pomeranz pays most serious attention to issues of gender. Pomeranz's concern to give due weight to the work and status of women in both Chinese and European paths to modernization echoes my own concerns and those of other feminist historians such as Susan Mann, but his questions about efficacy and choice are formulated principally in the terms of economic rationality. I feel that the nature and products of what I have called "womanly

71 Arrighi *et al.* (2003); Arrighi (2007); Chase-Dunn (2010) Elvin (2008) notes that he predicted precisely such a resurgence in *The Pattern of the Chinese Past.*

72 For example, Huang (1990); Brenner and Isett (2002). For a perceptive account of the debate, see Duchesne (2005).

73 Jack Goldstone, for instance, holds that China's "efflorescence" in the eighteenth century was an extraordinary achievement that deserves proper recognition (Goldstone, 2002).

74 The historical dynamics and typologies of wet-rice farming, and the consequent challenges to "Green Revolution", were one of my first projects in this vein (Bray, 1986).

work" merit closer attention if we wish to understand the full impact of China's long-term economic transformations (see Chapter 3).[75]

A related theme where my own work overlaps with that of the California School is the reconstruction of demographic mentalities. My research on reproductive techniques and choices in late imperial China in some ways confirms, and in some ways challenges the work of James Lee and his colleagues. They notably argue that the later Qing dynasty saw the emergence among important sectors of the population of modern attitudes towards optimal family size. While my own research on cultures of abortion has been cited by Lee and his colleagues in support of their arguments,[76] I include here an essay using medical cases, Chapter 5, which asks how well the evidence offered in these qualitative sources supports the claims that Lee and his colleagues build on quantitative data. In highlighting the uncertainties and anguish associated with problems of fertility, I illuminate some of the historiographical challenges involved in reconstructing demographic mentalities, or what I prefer to call reproductive cultures.

Lee's propositions about reproductive choice link to arguments that late imperial China saw the emergence of a consumer culture, and of social groups motivated by values that might be called middle class.[77] Since the social sciences took a consumption-focused turn a couple of decades ago, it has become common to treat certain levels and styles of consumption as catalysts of modern subjectivities and rationalities. In the process, such typical features of modern Western life as the concern for *privacy* acquire a new heuristic power. Modernization theorists have long argued that the consolidation in Western nations of legal rights to private property (including intellectual property as enshrined in patenting) was a key factor in the development of industrial capitalism and modern science. With the new, broader interest in what it means to have stuff, one theme that has attracted much attention is the relationship between modern individualism and the material and emotional experiences of spatial privacy – from the locked drawer to the gated community. In Chapter 4 I explore some spatial theories of privacy and modern subjectivity, to test how well they apply to some surprisingly similar material experiences and styles recorded in late imperial China (including, paradoxical as it may seem, to certain effects of female seclusion).

Among the iconoclasms of the California School, the most promising alternatives to Eurocentric narratives are in my opinion those proposed by R. Bin Wong. What

75 There are now a number of fascinating case studies in this vein. They include Fong (2004); Lu (2004); Huang (2012); Li (2012); and Tunstall (2012) on late imperial womanly work. Flitsch (2008); Wu (2008); and Jacob Eyferth's current research on 'Cotton, work and gender in Revolutionary China' examine the impact of technological changes on womanly work and the attributes of virtuous or successful womanhood in the rural PRC.

76 Bray (1997: 317–334).

77 For example, Brook (1998); Clunas (1999).

I find specially appealing is Wong's attempt to push historiography beyond 'the limits of European experience', in other words truly to take China 'on its own terms' rather than taking a Western genealogy as the pattern to measure how well the Chinese experience matched up.[78] Wong's strategy of "symmetrical comparison" between late eighteenth-century China and Europe proposes the equivalent of a Strong Programme for challenging the certainties and teleologies of eurocentric narratives.[79] Rather than comparing the performance of Europe and China from a modernist standpoint, instead Wong sets out to measure Europe against Chinese norms and goals of the period, and vice versa. Wong's study succeeds in developing an effectively non-teleological comparison of state-building and of regimes of governance.

In his challenge to the conventions of comparative history Wong does not, however, propose any critique of the prevailing historiography of technology. Beyond noting that explosions of change were not normal anywhere in the world until modern times, Wong's discussion of technology and economy, organized around a critique of Joel Mokyr's *Lever of Riches*, is perfectly conventional in its approach to which technologies counted and why on both sides of the globe (agriculture, transport, mining, textiles, metallurgy).[80] Yet one specially significant Chinese *sociotechnical system* leaps out of the pages of *China Transformed*, although it is not, of course, referred to in these terms: that is, state granaries. A central thread in Wong's argument is that the emerging polities of early-modern Europe and the state in late imperial China saw their rights to extract taxes, and their duties to protect their subjects from hunger, in very different lights – and that they put in place very different networks of institutions and technologies to pursue their ends.

While in *China Transformed* Wong does not address the technologies per se of "nourishing the people", he had already devoted an earlier work to the Qing public granary system and policies of price regulation and famine relief.[81] The statecraft tradition of using ever-normal granaries, *changping cang*, to balance surplus and dearth dates back to the Warring States, but there were continual changes in the networks of material, social and intellectual practice in which the granaries were embedded. How important were they considered at different periods? How effective did they prove? How were they built, organized, managed, filled and emptied? What alternative resources did the state rely upon or encourage to address the risk of hunger? The sociotechnical systems that coalesced around these granaries were a significant

78 Wong (1997); Elman (2005).

79 Rather than accepting the teleology that proposes that good science naturally drives out bad, the Strong Programme in the sociology of scientific knowledge, formulated in the UK in the 1970s, insisted on examining the success or failure of all knowledge claims, scientific or not, as historically and socially contingent processes.

80 Wong (1997: 55, 53–72).

81 Will and Wong (1991); see also Will (1990).

and persistent dimension of statecraft, but the key techniques, the resources and boundaries of the system at any point in time or space were distinctive, reflecting broader patterns in the economic organization and the working landscapes of late imperial China.

There were, for instance, recurrent debates from Song times onward over the merits of public, state-run granaries as against "charitable granaries", *yicang*, run by local gentry, or of simply relying upon markets to supply shortages and ensure fair prices. At one level these debates were highly ideological, but the practical, material implications (the challenges of management at different scales, the different kinds of building and means of transport, how long different types of grain would keep in different quantities) were also crucial factors in the arguments. Sometimes the state supplemented the granary system by investment in irrigation, or by opening up new lands to migration, as ways to spread the burden of feeding its people more evenly across its territory. During the early Qing dynasty a system was set up to report local rice yields and prices, so that officials could make more informed choices about when and how to intervene – despite doubts expressed at the time about how reliable the reports were, for well over a century the system was impressively effective.[82] In contrast, writing a few decades earlier when the fall of the Ming dynasty seemed well-nigh inevitable, the statesman Xu Guangqi noted that neither public nor private granaries were presently adequate to protect rural farmers against the recurrent threat of famine. Although he included valuable technical instructions for the financing, construction and maintenance of public granaries in his great opus 'Complete treatise on agricultural administration', *Nongzheng quanshu*, Xu Guangqi believed that, realistically, a more effective way to protect peasant farmers from the effects of crop failure was to teach them to grow nutritious and hardy crops such as sweet potato or turnips on a portion of their land, and to familiarize them with edible wild plants.[83]

The interest for the historian of observing the shifts over time of the imperial granary system is not so much that we can trace "progress" or "stagnation", but rather that it enriches our understanding of governance, its perceived challenges and the technical responses they elicited, at different points in time. There are, certainly, linear trends that strongly mark the unfolding of these patterns through the late imperial era. These include a steadily intensifying commercialization of the economy from the mid-sixteenth century; a new dynamic of unbroken population growth that began with the consolidation of Qing rule in the late seventeenth century; and irreversible environmental damage whose ravages became increasingly apparent across Chinese territories from the late eighteenth century. The governing classes acknowledged these unfolding trends and sought to address them, but they were not necessarily perceived as the most urgent or significant

82 See, for example, Will (1990); Marks (1998a, 1998b); Cheung (2008).
83 Bray (1984: 407–412); Bray and Métailié (2001).

issues of the time, nor as the unmistakable signals – as they seem so clearly to us now – of an irreversible transition into a new era.[84]

Court control over crafts is another case of a sociotechnical system whose shifts over time speak volumes about ideologies of rule, and about technological cultures. Ample materials exist to study this crucial technology of governance and the forms of efficacy attributed to it even in quite early periods, with more sources and materials continually coming to light, throwing the particularities of different dynasties or periods into illuminating relief.[85]

As a magnificent collection of new research at the Beijing Palace Museum demonstrates, the alien Qing court was determined to continue the long native tradition of central authority over the production of such material appurtenances of rule as fine silks, porcelain, bronzes, jade carvings, interior design and furnishings. The first Qing emperors chose to organize the control, production and circulation of the technical knowledge, skills and materials in quite novel ways, however, while adding new and politically charged crafts such as Western enamels and Tibetan bronze statues to the repertory.[86] Detailed documentation as well as models, sketches and the artefacts themselves testify that the Qing emperors did not merely delegate the task of developing technological developments in these crafts, but personally engaged in processes of experimentation and design.

As alien rulers, unable like their Ming predecessors to count on the cultural loyalty of the majority of their subjects, the Qing emperors deliberately engineered patterns of collaborative production to forge unifying bonds between the different territories, ethnicities and even religions of the empire – they developed these luxury crafts as technologies of nation-building in which material and symbolic efficacy were inextricably entwined.[87] Whereas early Ming regulations restricted the best

84 See, for example, Brook (1998) on changing attitudes towards commercialization, and Elvin and Liu (1998), Marks (1998a, 1998b), and Elvin (2004) on reactions to environmental degradation. While none of these phenomena were unprecedented, previously they had been repeatedly stemmed by wars and invasions, and/or by the introduction of strict new policies breaking with the legacy of the previous dynasty.

85 For example, Barbieri-Low (2007) on the Han dynasty; Kuhn (1987) on the Song.

86 See the essays in Schäfer (Xie Feng) (2010); Moll-Murata *et al.* (2005) is an invaluable complement to this research.

87 Cloisonné enamels, were European products that like other admired marvels of Western technical skill (watches, for instance) had hitherto never been matched in China. As for the Tantric bronze statues, after many centuries of tributary relations between Tibetan polities and Chinese dynasties, the Qing rulers from Kangxi onwards imposed closer control over Tibet and appropriated large swathes of its territories and peoples as part of their empire. The early Qing emperors furthermore patronized Tibetan Buddhism, and built Lamaist temples in the Chinese interior, as part of what is now considered a deliberate effort to construct Manchu identity as Inner Asian rather than Han Chinese (see Köhle, 2008 for a review of the debates). Not all the Qing imperial efforts to master and surpass other people's crafts were crowned with success; see, for example, Zhang Qiong (2010) on the mixed results of the early Manchu efforts to appropriate sericulture.

craftsmen to working exclusively for state workshops, the Qing rulers allowed them to apply their skills for personal profit too. This policy fostered both the circulation of technical expertise and the enhancement of luxury consumerism and production beyond the court. Yet this modernizing impact, as we might see it now, was quite incidental to the Qing emperors' agendas as they pondered on what kind of technical knowledge could be transmitted, how they wanted to transmit it, and to whom. Once again, studying the technologies from the perspective of "progress" would more likely hamper rather than heighten our understanding of the efficacy that was attributed to them at the time.

I have said little so far about *androtechnics*, the term I proposed for investigating how associations with technology shape types of masculine identity in a particular society. Are there specific technologies or sets of technologies that we can link to certain kinds of male role, either as ideal attributes or in actual practice? And how might such associations vary historically or otherwise? Beyond a few well-worn stereotypes (peasants with hoes and knights in armour, Man the Hunter and Man the Machine-Maker), this is hardly a well-developed area of historical research even in the West, and it is only now that a few sociologists and anthropologists are beginning to develop this approach for the contemporary industrial world. As the pioneering work of specialists such as Marie-Claude Mahias, Mareile Flitsch or Wu Xiujie demonstrates, analyzing gender regimes through technology, and technological cultures through gender, not only illuminates how other societies work, but calls into question many comfortable assumptions we make about our own, including what we see as natural or normal paths to modernity, and what we assume to be natural male or female skills or predispositions.[88]

My own research on gendered identities in late imperial China has focussed most obviously and specifically on femininities. But female implies male just as male implies female, and in mapping the terrain of idealized and actual femininities I inevitably pinpoint some key aspects of male identity too – whether rigidly orthodox, easy-going or subversive. Different registers or nuances of masculinity come through quite clearly in all the chapters on gynotechnics below, but perhaps most clearly in Chapter 1 on the ethics of domestic space, and Chapter 4 on forms of privacy. Much ink has flowed on how the ideals and realities of spatial seclusion shaped women's lives in imperial China. In contrast it sometimes seems as if Chinese men were unchallenged lords of space, free to stride or skip where they would. Both chapters explore the different kinds of "social space" available to men within their homes, noting that certain forms of spatial confinement, exclusion and regimentation were also part of masculine experience.

88 See, for example, Mahias (2002) on the commercialization of potting in a Nilgiri village; Flitsch (2008) on *kang*-centred skills; and Wu (2008) on successive phases of electrification in Chinese villages.

While I believe there is ample scope in imperial history for exploring the techniques of the male self, and technologies of the masculine subject, through a whole range of technical skills and material tasks or settings, it is within the context of the sociotechnical system of agriculture, including the technologies of literacy that were applied or developed in its service, that I have the best examples to offer. *Nong* denotes the peasant, the man who drives the plough, and his work, the tasks that he performs to produce food from the land.[89] By extension *nong* covers a whole range of activities and products necessary to the success of this seemingly humble enterprise, from irrigation to locust control, tax adjustments, the regulation of grain prices or the spring sacrifices by the emperor to the gods of grain and soil. My research on *nong* has necessarily addressed the role of the other iconic male technology, *shu*, the written document, in shaping *nong* as a system and giving it force as a technology of rule, as an ideology and as a science. Written documents were essential for both official and private record-keeping and administration; they were also an indispensable medium for codifying and transmitting technical knowledge, shaping the specific forms of agricultural science and the moral landscapes of the empire.

The spectrum of technical skills, knowledge, social identities and ethical commitments that collectively fell into the category of *nong*, agriculture, offer a clear case where a single technology was constitutive of several contrasting masculine identities, ideal and actual, complementary or contradictory. Farming in imperial China was unique in being a form of physical work approved for all social ranks: peasants, gentry, officials and even the emperor – driving a plough was honourable employment for all of them. Private landowners, scholars and ministers took pride in possessing the technical knowledge and experience needed to farm well. (So, I am sure, did peasants – but here as for crafts the feelings and ideas of the workers are not easy to retrieve except through attributions made by their literate social superiors.)[90]

Officials, however, operated at different scales from private landlords, and with different material and moral goals in mind: they were committed to the needs of

89 Although in practice women were likely to participate in a range of farming tasks, work in the fields, and ploughing especially, were conceptualized and represented as male activities; the essential complement was the iconically female, indoor activity of producing textiles. The gendered division of labour was routinely referred to through the binomes *nongsang* (farming and mulberries [for feeding silkworms]) and *gengzhi* (ploughing and weaving). These attributions are discussed in Chapters 2, 3 and 7.

90 Some historical forms of evidence about how craft-workers saw their skills have survived, including guild or craft stelae and some songs or religious rituals. A good example is the stela erected by a paper-making lineage in Sichuan, discussed in Eyferth (2009); see also Janousch (2011) on salt-temple stelae, and Cho (2008) on sericultural songs and temple fairs. Various ethnographic projects also illuminate worker skills and how practitioners foster and transmit them (e.g. Watabe *et al.*, 2003; Flitsch, 2008).

the state and the welfare of "the common people", *min*, whereas landowners were responsible to their own families. Referring back to the question of what kinds of knowledge were considered valuable, and to whom they should be circulated, one aspect of contrasting masculinities that I find fascinating is the tensions and differences between authorial personae reflected in the production of farming treatises, *nongshu*, composed by private landowners or by officials. In Chapters 6 and 7, I highlight epistemological, ethical and political distinctions between these two forms of knowledge production and circulation. Chapter 2 addresses how the forms of technical knowledge developed in China's earliest extant farming treatise, the *Qimin yaoshu* (Essential techniques for the common people) related to the mode of production within which the author, Jia Sixie, operated. Jia operated a centrally managed estate in North China, where millets and other dry-land crops were the staples. The best practices and economies of scale appropriate to this farming system operated in continual tension with prevailing statecraft ideals and policies for promoting peasant farming. But the imperatives of the Chinese agrarian state proved to mesh far more comfortably with the small-scale rice-based farming systems that formed the core of the imperial economy by late imperial times. Nevertheless, as I discuss in Section III, significant differences or even contradictions are apparent even in late imperial treatises between the goals and forms of agronomic knowledge developed in the service of the state, or by private landowners.

An interesting issue concerning historical narratives also arises here: to what extent can this type of natural knowledge formation, and its feedback loop with technical practice, be seen as "progressive"? Were Ming farming treatises better than their predecessors, and if so, how? In Chapters 6 and 7, I discuss the critical (or sometimes uncritical) processes that went into building up technical knowledge in different genres, looking in particular at crisis or locality as triggers for producing new forms of authoritative knowledge. In the history of European science, the development of technical drawing is heralded as a crucial breakthrough in ways of thinking analytically and creatively about technology. In China various forms of technical drawing were developed, and among those that circulated most widely were the illustrations to agricultural works – but by the end of the Ming dynasty the enterprise of making new agricultural drawings, with one or two exceptions, had simply petered out. What are we to make of this?[91] In Chapter 8, I look at the particular case of illustrations, comparing the forms, significance and historical trajectories of two distinctive traditions of depicting agricultural techniques and implements.

91 On the significance attributed to technical drawing in Europe, see Ferguson (1977) and Lefèvre (2004). For a range of reflections on the place, potential and limitations of technical drawing of various kinds in late imperial China, see Bray (2007a), Golas (2007), and the papers in Section III of Bray *et al.* (2007).

One of these traditions is embodied in the charming scenes of the *Gengzhi tu*, and at this point I return to the figures with which I opened this Introduction. How do we explain that three of the most energetic and powerful rulers of the Qing dynasty, all dedicated to effective government and anxious to secure the prosperity of their rural subjects, each invested time, money and energy in producing beautiful and elaborate new versions of the twelfth-century original, none of which gave any hint of the technical changes that had occurred over the intervening centuries? In brief, the answer lies in the complexities of *quannong*, "promoting agriculture", a moral commitment for every Chinese ruler, and an endeavour that invariably interwove material and symbolic support.

Historians of technology have usually presumed that the *Gengzhi tu* was originally designed to provide advanced technical knowledge that could be used to improve agricultural and textile practices. Although Lou Shou's original pictures were admired by contemporaries for their technical accuracy and detail, however, there is no evidence that they were ever used for technical instruction. In terms of *quannong*, the efficacy of the scrolls lay as much in the tenor of Lou's poems, poignant evocations of the exhausting labour that farming families endured, and of the sacrifice, dedication and skill that generated the country's wealth. For the Qing emperors, who were certainly no laggards when it came to practical measures for increasing agricultural output,[92] the value of the *Gengzhi tu* lay not in its potential for technical pedagogy, but in its iconic powers: it symbolized an ideal social compact, a bond between the ruler and his peasant subjects that was both material and mystical. In promoting a *Gengzhi tu* revival, the Manchu rulers presented themselves as legitimate heirs to the Confucian tradition of statecraft and political ethics. For ordinary people, meanwhile, scenes from the *Gengzhi tu* celebrated a fruitful social and cosmic harmony, the essential complementarity between male and female work, the dignity of ordinary work and its rewards. The central theme of families working together to achieve abundance is surely one reason for the work's popularity in late imperial times and through into the twentieth century: these scenes ranked with depictions of rosy-cheeked infants at play as auspicious images of happiness, virtue and harmony.

The section on rice farming in the original version of the *Gengzhi tu* ended with the filling of the granary. The version commissioned by the Kangxi emperor in 1696 added a new and suitably solemn scene of a thanksgiving ceremony at the family altar (Figure 0.2). But a more popular version from a few decades earlier gave a very different emotional twist to the significance of the set of pictures: it added a picture of a riotous harvest feast at the landlord's house, with drunken tricks, dancing and much vomiting (Figure 0.3).

92 See, for example, Will (1994); Rowe (2001); McMahon (2009).

FIGURE 0.2 'Thanking the gods', *Gengzhi tu*, 1742 woodblock version based on Jiao Bingzhen's painting of 1696.

I have said much in this Introduction about technology and power, about discipline, about morality, about the state. Readers may have concluded that the following chapters offer only pictures of domination, toil and unrelenting gloom. But my sources also offer glimpses of ingenuity, leisure and enjoyment, pride in work well done and appreciation of skills and beauty. I hope this book may inspire others to take on technology as a serious theme, not only to bring into their historical projects the sad and serious mechanisms of toil and subjugation, but also to look more closely at what a technological perspective can add to our historical understanding of life's everyday satisfactions and pleasures. My more serious goal, however, is to rehabilitate technology as a seductive theme for historians, an integral component of everyday life, of statecraft and of knowledge whose workings we have only begun to explore.

FIGURE 0.3 'Farmers' celebration', *Bianmin tuzuan*, 1593 edition.

SECTION I

Material foundations of the moral order

This section considers the mute yet powerful role of material artefacts and environments in choreographing everyday behaviour into moral patterns. The two cases discussed are the place of the ancestral shrine in domestic architecture, and the historical evolution of the real and idealized farming landscapes that constituted the base of the agrarian order. Both sociotechnical systems began to take shape in the Song dynasty. Like cathedrals in medieval France, or the automobile in our modern world, they shaped society in ways that far transcended the material. As they developed over the ensuing centuries, housescape and landscape alike came to constitute shared frames of moral experience for the ruling elite and ordinary families, while at the same time offering sufficient material and interpretative flexibility to accommodate the changing exigencies of everyday life and material survival. In both cases it was the ambiguities and adaptability of these moral-material matrices, I argue, that gave them historical significance and staying power as stabilizers of the imperial order through times of crisis and change.

1

MACHINES FOR LIVING

Domestic architecture and the engineering of the social order in late imperial China

In about 1169 the philosopher Zhu Xi set down a series of recommendations on the construction of an ancestral shrine in what was to become his best-known work, *Family rituals*. The first chapter began as follows:

> When a man of virtue builds a house his first task is always to set up an offering hall [ancestral shrine] to the east of the main room of his house.[1]

This passage has not usually been thought of as a key document for the history of Chinese technology. I should like to propose it as such, arguing that in the context of Chinese society of the late imperial period (between about 1000 and 1800), domestic architecture was a technology of significance comparable to that of machine-tool design in the nineteenth-century United States: the shrine was a core element in the formation of a pervasive, flexible and enduring sociotechnical system,[2] a material artefact around which crystallized a characteristic ideology and social order. By taking the construction of ancestral shrines rather than the production of guns or steel as my subject, I hope to suggest how history of technology can be enriched and connected to other fields of history from which it is at present all too frequently estranged.

1 *Zhuzi jiali* ('Family rituals of Master Zhu', tr. Ebrey, 1991a: 5). In fact, this sentence was quoted from a handbook on etiquette published a century earlier by the statesman Sima Guang (see below). Zhu was most famous for his numerous scholarly works on moral philosophy and cosmology, but none reached as wide an audience as *Family rituals*.

2 'The distinctive technological activity that stems from the linkage of techniques and material culture to the social coordination of labor' (Pfaffenberger, 1992: 497).

Historians of technology have usually considered that truly significant technologies are those that catalyze radical transformations. They give very high marks for change and low marks for continuity (often labelled "stagnation"). Comparing the technological dynamism of early-modern Europe to China at the same period, the questions they ask have all too often been: what was missing? What went wrong? Yet the goals of the Chinese state – as of most governments past and present, including those of the modern West – were to maintain the social and political order, often in the face of almost overwhelming pressures for disintegration. Between 1100 and 1700, China was twice conquered by foreign armies; its population quadrupled; the state fought against pirates and rebels, and struggled to protect its population against floods, droughts and famines. Commerce developed dramatically, and a largely subsistence economy was transformed into networks of provincial, national and international trade. In the early part of the period an aristocratic elite was replaced by an educated meritocracy of landed gentry; by 1500 the gentry were fighting against the pretensions of merchant families buying or studying their way into the elite. Technologies of various kinds served to counter these fissile tendencies. Some were consciously viewed in this way and deliberately deployed by the state and its officials. Water control, state printing houses, communal granaries, roads and canals, and of course the army are examples. At a more everyday level, the material design of the house, and the normalization of domestic spatial practices, effectively worked to draw increasing numbers of Chinese families into a shared social order organized around the patriarchal lineage. Although the Chinese state did not directly intervene in domestic construction, it encouraged the development of a shared habitus of respectability in numerous ways, so that across the social spectrum Chinese houses became machines for living according to a characteristic set of ethical principles.

Machines for living

The American System of Manufacture fundamentally reshaped production and laid the foundations for the modern industrial world. It had its roots in "armoury practice", a system of standardization of machine tools specially designed to produce large numbers of interchangeable parts, which started in the US small-arms industry in the early nineteenth century and spread rapidly into other areas of manufacture. In Europe, skilled craftsmen adjusted their machines to produce components of the required design and dimensions; to profit from an influx of unskilled immigrant labour, US manufacturers designed machines where the craftsman's skills and tolerances were in-built. These machines propelled the development of a new manufacturing system, which engendered the labour relations, consumption patterns and vision of technology typical of Fordist industrial capitalism.[3] The *normalization*

3 See, for instance, Pursell (1994).

of industry crystallized around 'the U.S. Ordnance Department's 1816 commitment to the philosophical ideal of standardization and interchangeability', and gradually came to encompass not only the structure of manufacture, but also 'the growth of standardized and centrally controlled rail systems, the centralization and standardization of corporate research and development, the use of consumer advertising to program individual buying habits, [and] the increasing centralization and complexity of electricity and communication networks'.[4] In other words the machine-tool came to underpin a whole system of goals and values – it was not just a machine for producing, but a machine for living.

When identifying *significant* technologies, those that have contributed most to shaping the nature of society, historians of Chinese technology have usually followed the example of Western historians and focused on technologies that produced the key commodities of the industrial world – metallurgy, agriculture and the textile industry. However, late imperial Chinese society was not capitalist, and its characteristic social order was organized around other than modernist goals and values.[5] The institution that most fundamentally shaped late imperial society and culture was the patriarchal lineage, constructed around an architectural feature: the domestic altar. In China, domestic architecture was a *significant technology* that *normalized* late imperial society just as the machine-tool shaped and cemented the values of the Fordist world.

In speaking of his architectural designs as machines for living, Le Corbusier expressed the modernist ideal that domestic space should be designed to meet scientifically defined needs as efficiently as possible: a modern dwelling should be hygienic, thermally stable, efficient in its use of energy, and ergonomically designed. The phrase also expressed the converse modernist principle, that living spaces produce lifestyles and subjectivities. Bauhaus workers' flats reduced drudgery and increased comfort – and they were also designed as environments that would help the occupants live as responsible modern worker-citizens. In contemporary Western society new housing is professionally designed, incorporating the skills of architects and engineers. The building materials are industrially produced. The house itself and its standard equipment tie the occupants into a complex technological network of utilities suppliers, consumer durables industries, roads and automobile manufacturers, supermarkets, TV cable companies, and so on. The house is a machine for living a particular lifestyle, that of late capitalism with its characteristic systems of consumerism, property rights, privacy, and gendered or generational identities. Excellent studies by Dolores Hayden and Mike Davis illuminate the moral, cultural and political messages embodied in such dwelling forms as the "executive housing"

4 Staudenmaier (1985: 200).

5 Not to say that seeking for profit was unknown, far from it – but it was a "muted" discourse in late imperial society (e.g. Brook, 1995: 84–90; Chapters 6 and 7).

now spreading from the United States to middle classes around the world, and invite us to ponder the deeper meanings of gated estates, of house frontages giving pride of place to two- or even three-car garages, the absence of shared facilities with neighbours, a separate but tiny bedroom for each child, and, of course, multiple bathrooms.[6]

We have naturalized the moral and cultural messages of this architecture. We think of a separate bedroom for each child, or a closed-off space for private, solitary defecation and ablutions, as universal needs that all societies would meet if they had the means. However, as anthropologists and cultural critics have shown, architecture is not neutral.[7] A house is a cultural template; living in it inculcates fundamental knowledge, skills and values specific to that society. It is a learning device, a mechanism that converts ritual, political and cosmological relationships into daily experiences of space.[8] The encoded messages of the house teach some lessons that are the same for all, and some that are different. As a child grows up and learns the practices of living in a house, she learns her proper place within society; she internalizes the hierarchies of gender, generation and rank that are marked by walls and stairs and practised in the rules and etiquette of receiving guests, performing rites of passage, and going about daily tasks. She learns about respectability, and how the differences between high and low, rich and poor are concretely expressed.

Much of this cultural expertise forms around material objects, skills and habits. Within the framework of our own cultural expectations the *technical hardware* of modern domestic architecture "dedicates" certain spaces, restricting their use. When we move into a new house we do not contemplate setting up a dining table in the bathroom, nor would we consider relieving ourselves in the bedroom now that the flush toilet has replaced the chamber pot in our lives. But these uses of space and of equipment are neither self-evident nor "natural" – they have to be learned. It is in the nature of capitalist society that new features (the fitted kitchen, the "study") are constantly being developed and presented to consumers as needs, and in such cases we find ourselves instructed and enticed by a range of *users' guides* that may include floor plans in the realtor's window, magazines on furnishing and

6 Dolores Hayden (1986) discusses the ideology of American middle-class housing design from a feminist perspective. Mike Davis (1992) and Setha Low (2003) focus on how contemporary architectural design and its emphasis on security reinforce class difference and segregation.

7 Siegfried Giedion (1948) provides wonderful insights into the machine-age ideology inherent in such apparently culture-free concepts as "hygiene" and "comfort".

8 The classic text on house and habitus is Pierre Bourdieu, 'The Berber house' (1973). Comparing sedentary and hunter-gatherer cultures, Peter J. Wilson (1988) argues that the material experiences of inhabiting dwellings have profoundly reshaped our cognitive patterns and social values, indeed that we owe such institutions as "the family" to the spatial experiences of the house. See also Carsten and Hugh-Jones (1995) for an exploration of the connections between house and family structure.

design, or TV advertisements. Occupants draw upon these as well as their own experience and expectations in order to turn four walls into a home.

Engineering a new social order in late imperial China

In what follows I describe the interplay between *hardware* (that is to say the characteristic architectural features that framed spatial practices) and *users' guides* (texts and shared practices that taught or reminded people what the spaces they occupied signified) in the construction and diffusion of a "standard" Chinese house. The normalization of domestic space, I argue, was a key element in the standardization of social practices and values.

In his superb study of the Chinese carpenter and his technical skills, Klaas Ruitenbeek makes an arresting observation: 'One gets the impression that in China *an imaginary architecture existed which was superimposed on ordinary architecture, and which was the primary concern of the owner, geomancer and carpenter alike*'.[9] Ruitenbeek is referring to the superimposition on the physical building of an energetic (geomantic) structure (see below). In my own study of domestic architecture and the construction of gender I eventually came to see three "imaginary architectures" superimposed upon the material shell of the house, each conveying a different set of messages about the relations between its inhabitants, the cosmos and society at large. First, the Chinese house was a space of decorum, an embodiment of orthodox neo-Confucian values; second, as Ruitenbeek and many anthropologists have demonstrated, the house was a cosmic, energetic space; and third, it was a space of culture, representing a Chinese view of what home and shelter should be.[10] In the following analysis of the diffusion and reception of the shrine, I concentrate on the interplay between the material structures of the house and the two "imaginary architectures" that produced the space of decorum and the energetic space.

While details of the structure and use of domestic dwellings in China obviously varied according to period, region, class, wealth and taste, what I call the "standard" house, a concrete material embodiment of neo-Confucian social values, was first defined in texts written in the Song dynasty (960–1279). At the beginning of the dynasty only a few gentry families were claiming for themselves the aristocratic privilege of an ancestral shrine in their homes, but in order to consolidate their

9 Ruitenbeek (1993: 62, emphasis added).

10 Again, while not wishing to essentialize "Chineseness", and despite the range of variation across China, we can legitimately point to obvious shared aesthetics, preferences for particular construction materials, the absence of open hearths, the understanding that the house should be constructed and decorated in such a way as to bring good fortune to the family. See, for example, Lu Yuanting (1991–1996); Bray (1997: Chapters 1–3); Knapp (1998); Knapp and Lo (2005).

own status as social leaders the emerging neo-Confucian elite encouraged ever-larger segments of society to adopt the same configurations of spatial layout and practice in their own homes, along with the social ties and values that they symbolized. By the nineteenth century, the universal feature of Chinese – and indeed Korean – houses that immediately impressed itself upon Western visitors was that they all, however grand or humble, were designed and organized first and foremost as ancestral temples.[11]

Until about 1000 CE the ruling elite of China had consisted principally of aristocratic families who preserved their own culture and privileges by *excluding* other social groups; in particular they reserved for themselves the right to domestic rituals of ancestral worship. Thereafer, a new elite became predominant, formed of public servants qualified by education, not by blood. Rather than drawing impermeable boundaries around themselves, the new elite saw themselves as a meritocracy open to talent. They developed and propagated a social philosophy and accompanying social practices usually referred to in English as neo-Confucianism. The roots of neo-Confucianism went back to the late Tang dynasty, to the eighth and ninth centuries, when the formalization of an examination system for the civil service allowed non-noble scholar-bureaucrats to undermine the political and cultural supremacy of the aristocratic elite. Classic Confucian philosophy had emphasized the organic nature of the state, the continuities between the regulation of the family and that of the polity, and the key importance of ritual in ordering society. Debates on these themes resurfaced in the Tang, sharpened by the threat to orthodoxy represented at the time by the "alien" doctrines and practices of Buddhism. By the early Song, commoner scholar-bureaucrats had definitively displaced the aristocracy in government, and the new forms of Confucian orthodoxy they developed justified and extended their own claims to social and cultural leadership, placing great stress on ritual.[12]

The neo-Confucian social order rested on principles of ranking and complementarity. A subject owed respect and obedience to his or her ruler, who should respond with consideration and compassion. Within the patriline the lineage head occupied an analogous position with respect to junior lineage members, and the relations between father and sons, husband and wife were similarly conceived as complementary but not equal. The performance of joint rituals served to unite people of different status in a common goal or set of beliefs while reaffirming the proper social hierarchies.

It has frequently been observed that the late imperial ruling class preferred to use instruction rather than force as a method of social control. Chinese thought emphasized the connection between body and morality; the physical performance

11 Clément *et al.* (1987).

12 See, for example, Kuhn (2009).

of daily rituals inculcated the proper feelings and appreciation of relationships that the rituals celebrated. Instructing the common people in ritual, wrote the statesman Ouyang Xiu (1001–1072), 'not only would prevent disorder but also would teach them to distinguish superior and inferior, old and young, and the ethics of social relations'[13] – note that in this formulation ritual has an active rather than simply a symbolic role: it *secures political order* by reinforcing social relations. It has been said that Chinese authorities were more concerned with orthopraxy (correct practice) than with orthodoxy (correct belief), since where practice led, belief must follow.[14] Hence the moral significance of bodily practices, including work, and of material artefacts: everyone knew, for example, that the philosopher Mencius (fourth century BCE) had announced his brilliant moral career as a mere infant by playing with sacrificial vessels long before he could lisp their names.

As Foucault has reminded us, everyday artefacts and practices are the more powerful as disciplinary devices because their messages are silent. Domestic architecture was a particularly powerful instrument of moral inculcation in China because of the integration of private and public ethics characteristic of late imperial society. In the West we have experienced a series of separations between spheres of activity that has progressively reduced the dominance of the house in shaping people's lives. The Greek and Roman worlds already marked a clear separation between the domestic and the public, political domain; in Christian Europe most important acts of worship and rites of passage came to be performed in the church; and with industrialization came a separation of workplace from dwelling for large numbers of the population. The house became a private domain, a refuge from the pressures of work and of political or religious orthodoxy. In China a contrary historical process occurred, whereby an increasing proportion of the population came to be tied into a polity that did not recognize our distinctions between public and private: rituals of birth, maturation, marriage and death were performed in the house; the liturgical rituals of the domestic ancestral cult paralleled the ceremonies of state orthodoxy; and the family was the ethical and behavioural training ground for political life.

Nineteenth- and twentieth-century Western observers remarked that every Chinese house, whether of peasant or gentleman, was first and foremost an ancestral temple: the entire structure was centred on the shrine, and when a household divided, each brother set up an altar of his own in his new dwelling.[15] Though it seemed that such customs were immemorial, in fact peasants – and even scholars – had not always been entitled to their own ancestral shrine.

13 de Bary (1960: 443).

14 Watson (1988).

15 See, for instance, the classic account by Myron Cohen (1976).

Before the Song, the domestic altar with its ancestral tablets was an exclusive emblem of aristocratic status. Commoners were not allowed to worship their ancestors in their own homes. However, in the early Song dynasty neo-Confucian gentry families began constructing patrilineal descent groups and setting up shrines for themselves as marks of their elite status.[16] Domestic shrines (the architectural *hardware*) and the practices associated with them (described in various *users' guides*) served this new elite as an effective tool in their construction of a stable and integrated social order with themselves at the top.

In 1169, Zhu Xi, one of the most eminent and influential founders of neo-Confucian thought, published a compendium of domestic rituals for use among gentry families, the *Family Rituals*, incorporating the *Miscellaneous etiquette for family life*, composed a century earlier by Sima Guang (1019–1086). The combined text constitutes the foundational formulation of the "standard" spatial configuration of domestic space and practices.[17]

Zhu Xi represented the house as a ritual space with ancestral shrine at its heart: 'When a man of virtue builds a house his first task is always to set up an offering hall to the east of the main room of his house'. Zhu Xi stated that this section came first in his book because it was fundamental to all that followed, not only in moral and metaphysical terms but also for the inculcation of proper deportment: the first chapter, in his words:

> provides the basis for understanding the fine points in the later chapters concerning movements and postures, for walking here and there, getting up and down, going in and out, and facing various directions.[18]

Zhu Xi specified the general principles of construction of the shrine, but made allowances for disparities of wealth. Ideally it should be three bays wide, with chests to keep family genealogies and ritual utensils, but poor families could make a shrine just a single bay wide,[19] or even just use the east end of the main building. Zhu Xi also made allowances for other material inadequacies, such as the lack of a south-facing main building:

> Here and throughout this book, in organizing the room, *no matter which direction it actually faces*, treat the front as south, the rear as north, the left as east, and the right as west.[20]

16 Kuhn (1987); Ebrey (1991b).

17 One fundamental element of Chinese domestic spatial norms, the segregation of the sexes, will be discussed in Chapter 4.

18 Tr. Ebrey (1991a: 8).

19 A bay is the space between two pillars supporting a roof beam, about 10 or 12 feet across.

20 Ebrey (1991a: 8, emphasis added).

The absolute requirements of orientation were thus reformulated as a *set of transformations* such that anyone could conform to them. The centre of the room was the altar table and the niche containing the ancestral tablets – again, considerable variation was possible in the number of tablets and the order in which they were arranged; the main thing was that there *should be* an order.[21]

The daily offerings made to the ancestral tablets and the choreography of the liturgy, 'walking here and there, getting up and down', embodied the principles of filiality, ranking, and harmonious and fecund complementarity, which underpinned the neo-Confucian patrilineal and patriarchal order: the respect and obedience due from descendants to ancestors was the pattern of filiality, *xiao*, and thus the model for relations between children and parents, wife and husband, subject and ruler. One very important feature of shrine construction was the east and west steps of the room, illustrated clearly in Figure 1.1. Family members ascended and descended the steps in ranked order, according to generation, birth order within the generation, and sex, passing from the daily world up to the holy level of the ancestors and back; men and women were distinguished by using the eastern and

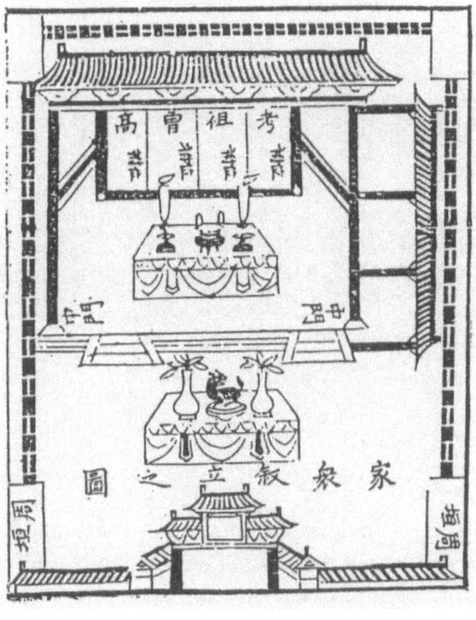

FIGURE 1.1 Family offering hall, from the 1602 edition of *Zhuzi jiali*. The characters on the screen at the back of the chamber indicate the genealogical order in which the tablets should be arranged. Note the two sets of steps leading up to the chamber from the courtyard.

21 For example Watson and Rawski (1988: 498) shows a Qing-dynasty merchant-family shrine containing five, not four sets of tablets.

the western steps. Here let us note that the basic unit in the liturgy (essentially a reproductive activity) was the *married couple*: every act a man performed (such as offering wine) must be matched by a complementary act by his wife (offering tea).[22] Family ceremonies reinforced patrilineal descent principles both by their choreography and by their inclusions and exclusions (Figure 1.2): married sons (who lived in their parents' house) participated with their wives and children; married daughters (who joined their husband's lineage on marriage) were absent; servants were not involved, nor were concubines, who unlike legal wives were not presented to the ancestors when they came into the family.[23]

According to neo-Confucian precepts, all family life and events should be organized round the altar: the ancestors must be informed of comings and goings and of success and failure, new brides were presented to them, dying family members

FIGURE 1.2 Celebration of an ancestral sacrifice, *Shinzoku kibun*. This illustration, composed c. 1800, shows that the sacrifice is performed by all the couples in the family, each wife standing behind her husband, and also by the unmarried sons.

22 On gender complementarity, patriarchal control and the different meanings that seclusion of women could assume once productive work such as weaving or reproductive work such as the raising of children were taken into account, see Bray (1997), especially Chapters 2 and 3.

23 Concubines never became part of their husband's lineage. The children to whom they gave birth were considered the offspring of the legal wife, and they were never allowed to conduct ancestral rites even if the wife died. In that case, the husband's duty was to remarry (Bray, 1997: Chapters 8, 9).

were set down beside the shrine to breathe their last. The senior couple in the family occupied the room closest to the shrine, as befitted both their rank within the family, and their closeness to death and ancestral status. If robbers broke in or there was a fire, the first things you should save, said Zhu Xi, were the ancestral tablets and family documents, leaving jewels and money until later (one suspects that this injunction was often ignored in real life).[24]

Sima Guang, writing in the eleventh century, appears to have been the first person to suggest that the practice of constructing one's house around the family shrine should become general among the literati elite. A century later, Zhu Xi took a much more radical step: his goal was to extend the privileges of lineage rituals to whole agnatic groups. He wrote his *Family rituals*, intended as a manual for local magistrates to educate the people in their charge, in a conscious effort to popularize, to provide a clear and practicable set of liturgical practices for general use – though it is probable that Zhu's goal was to educate families of standing rather than the masses.

In placing such emphasis on the shrine Zhu Xi was not innovating (after all, he begins by quoting from a text a century old) but trying to systematize a practice that had been spreading among the non-aristocratic gentry in the eleventh and twelfth centuries. The established elite disseminated the shrine and its associated neo-Confucian practices as they extended their control over local society, tying their poorer kin into the formal institutions of patrilineal descent groups; meanwhile other families anxious for elite status laid claim to lineages and kinship networks of their own.[25] Joint ceremonies honouring the founding ancestor took place at special lineage temples that were shrines writ large; they confirmed the landowning and educated elite as lineage heads and as ritual experts and advisers. Though ranked, the system was inclusive: if rich and poor were to be tied together in loyalty as descendants of a common ancestor, then poor families had to have descent (and ancestors, and daily rituals at family shrines) too.

Since Zhu Xi's liturgy allowed ceremonies to be carried out with propriety with the most basic of material hardware, poor families could achieve this significant form of respectability at low cost – thanks to the transformations suggested by Zhu Xi, a simple niche in the wall sufficed as the organizing device for the liturgy.

The influence of Zhu Xi's formulation of the "standard house" and its spatial practices and meanings spread like ripples in a pool. One source of ritual and spatial instruction for ordinary families consisted of the joint lineage ceremonies; for important domestic rituals such as weddings junior families would often invite a

24 Ebrey (1991a: 5).

25 Lineage-building was common right through the late imperial period: upwardly mobile local families would identify or invent a famous ancestor and compose advantageous genealogies for themselves; see the classic study by Rubie S. Watson (1985).

related gentry member to participate as sponsor. Another was the actual construction of houses. As apprentices, the carpenters who were principally responsible for building the wooden houses of late imperial China learned by heart the principles of construction enshrined in their sacred book, the *Lu Ban jing* (the Canon of Lu Ban [patron saint of carpenters], or Carpenter's canon). These included rules and measurements for the orientation, proportions, and dimensions of the hall or room housing the shrine, of the shrine itself, and of other key domestic features such as the marital bed. In the process of building a new house, or adding to an old one, a family went through a period of conscious concern with such issues.[26]

Zhu's book itself became very popular and was produced in various editions. The first illustrated versions appeared at the end of the Song (Figure 1.1 is from a Ming edition). The original text was popularized and excerpted in household encyclopedias, for example in the *Jujia biyong shilei quanji* (Householder's vademecum).[27] Although only well-to-do families could afford printed books until about 1500, thereafter the reading public grew rapidly, especially in the cities. The middle class, which developed between 1500 and 1800, liked to buy books, both to learn from them and as a sign of status, and books on ritual were very popular.[28]

As a concern with proper domestic ritual spread both socially and geographically, along with the spread of lineages, neo-Confucian thinkers became increasingly preoccupied with the problems posed by "vulgar practices" and "local customs"; they wrote commentaries on Zhu Xi's work, or general reconsiderations of ritual, that tried to negotiate these problems so as to include even greater numbers within the circle of orthodoxy. State techniques for extending orthodoxy frequently overlapped or coincided with gentry initiatives. They included legislation on ritual conformity (for instance, the Yuan dynasty [1279–1368] code laid down that only marriages conducted according to the *Family rituals* would be considered legal),[29] the setting up of schools, the organization of lectures (especially popular in the early Qing dynasty, during the eighteenth century),[30] and the granting of honours to subjects of outstanding virtue and merit.

It would be wrong to construe this relationship between state or elite and ordinary people as straightforward domination. Ritual and spatial conformity constituted a desirable and attainable sign of respectability. Norbert Elias decoded the spatial structures of the Parisian *hôtel* as a carefully designed stage on which the French aristocracy under Louis XIV gave uninhibited performances of what they saw as

26 Ruitenbeek (1993) addresses the history, contents and usage of the *Carpenter's canon*; see also Bray (1997: 159–166).

27 First published in 1301 and widely circulated in the Ming edition of 1560, revised by Tian Rucheng.

28 See, for instance, Chia (2002); Brokaw and Chow (2005).

29 Ebrey (1991b: 151).

30 Mair (1985).

the role of their class; its finely tuned aesthetics, relationships and forms of rationality were thrown into relief by contrast with the emergent bourgeoisie. Only fellow aristocrats could judge or join in these performances; others were either too vulgar (the bourgeois, confined to the role of envious onlookers) or not really human (the servants – thus a French noblewoman could undress without embarrassment in the presence of a male servant).[31] The lineage temple was likewise a stage designed by the neo-Confucian elite to display their own moral values and social pre-eminence. But this particular form of social pre-eminence required the participation of lower-ranked groups who shared and enacted the same values. The neo-Confucian elite invited their social inferiors not only to join them as actors on the stage of the temple, but to reproduce the performance in their own homes.

The Chinese elite saw their task as drawing the lower orders firmly into the domain of civility, while maintaining the distinctions of rank and learning that justified their own authority. In setting up their own domestic shrines, ordinary families recognized this social order but stood to gain respectability; ancestors; membership of the lineage; control over women (an attraction for men) and over juniors (also an attraction for senior women); and a sense of political incorporation. In neo-Confucian doctrine the house was a microcosm of the state, and relations between husband and wife paralleled those between emperor and minister. As the spatial norms of the house diffused through Chinese society in the course of the late imperial period, the daily rituals around the altar came to tie increasing numbers of ordinary and even illiterate families into an ordered social space stretching geographically to the bounds of the Chinese polity, and historically to the great scholar and moralist Zhu Xi, and so back to Confucius himself.

Material and moral ambiguities

All I have said so far suggests that the family altar was instrumental in propagating hegemonic, orthodox values. There is another side to the coin, however. One important factor in the popularization of the altar was a typically Chinese process of accommodation between "orthodoxy" and "popular custom" – *the geomancy of the house*, which modulated the decorous symbolism of neo-Confucian liturgy, superimposing its own rather different morality upon the family shrine.

Geomancy was applied cosmology; its skills consisted of *reading* a local landscape in terms of flows of cosmic energy (*qi*) through time and space, and then *manipulating* the landscape so as to direct energy in the desired direction. It was an ancient science that entered a new phase of elaboration with the surge of cosmological studies in the Song dynasty; only scholars and specialists had expert knowledge of the cosmological principles involved, but almost everyone had a

31 'Structures et signification de l'habitat', in Elias (1985 [1933]).

smattering of geomantic lore and was familiar at some level with the general precepts concerning the siting of graves and the layout of dwellings.[32] Carpenters, who were the craftsmen responsible for building the wooden houses typical of China, were practically speaking geomantic experts: the carpenter's rule was marked with lucky and unlucky inches, and when he calculated the width of a lintel or the length of a beam, he was carrying out a series of cosmological computations. Anyone building or altering a house, however poor, first consulted a geomancer – and if they could not afford a specialist, the carpenter who did the building work was sufficiently expert with a compass to stand in.[33]

As in neo-Confucian liturgy, the altar was the heart of the geomantic house.[34] After the most auspicious site for the altar had been located, the rest of the house was designed around it in a configuration intended to channel *qi* into the main hall where the altar stood. In the siting and design of a house, geomancy worked to direct *qi* to the advantage of its occupants. A well-sited and well-laid-out house brought health, wealth and happiness, and numerous male heirs; a badly sited dwelling brought strife between father and son, shameless daughters, loss of wealth, and illness (Figure 1.3).

Neo-Confucian philosophy said that virtue brings happiness and order brings success; everyday rituals around the shrine were a training in morality and an affirmation of social structure. Differences in roof height between the wings and buildings of a house, for instance, *reflected* difference in rank between occupants but did not *produce* it. Geomancy, however, said that the manipulation of *qi* through architecture *produced* virtue and proper relationships.

> A family temple is not like an ordinary house: whether or not sons or brothers will attain wisdom depends wholly on this place. Moreover the rear hall, main hall, corridors and triple gate may increase only gradually in height, since only then do sons and grandsons know their rank; and does not the younger aspire to the older's place. The builder must take careful notice of this.[35]

Geomancy was not necessarily incompatible with Confucian values yet, as Stefan Feuchtwang underlines, its underlying principles were asocial or even amoral.[36] Geomantic techniques were competitive: they did not increase the cosmic energy

32 Bennett (1978). There is a huge Chinese-language literature on housing geomancy; Knapp (1998) cites the most important works in his bibliography.

33 Ruitenbeek (1993: 6).

34 The siting of the kitchen stove was also extremely important and there are good grounds for thinking of the Chinese house as a bipolar construction in which the stove was as important as the altar in constituting group identity and its reproduction (Bray, 1997: 106–114).

35 A Yuan (1279–1368) passage from the *Carpenter's canon*, tr. Ruitenbeek (1993: 197).

36 Feuchtwang (1974).

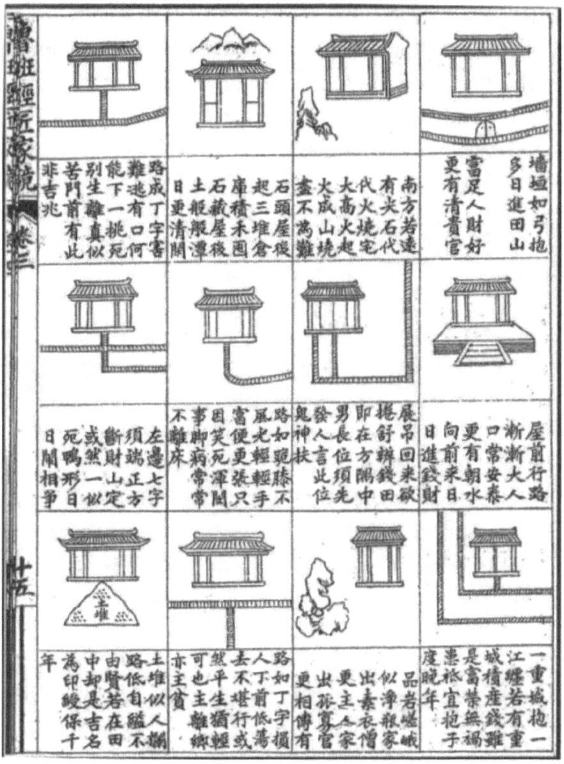

FIGURE 1.3 Page of diagrams from an 1808 edition of the *Lu Ban jing* depicting lucky and unlucky building forms.

within a local landscape, they simply channelled it in new directions. Siting a house or a grave, building a wall or even planting a tree disrupted existing flows of *qi*, and were regarded by the community as a means of attracting fortune for one's own house at the expense of others. 'Let one man in a village build a fraction too high; let him build a window or a door which can be interpreted as a threat; and he has a struggle on his hands.'[37]

In the cultivated neo-Confucian, reading the ancestral cult was the paramount symbol and instrument of social harmony and political order; the shrine was an object of respect, the focal point around which the performance of family virtues was organized; the ancestral tablets were the symbols of patrilineal descent. However, most people considered that their ancestors' spirits actually inhabited the tablets and intervened actively in their lives. The ancestors' power to help their descendants

37 Freedman (1969: 14).

was affected by the supply of *qi* channelled on to the altar by geomantic means – the shrine was in that sense a machine for converting cosmic energy into human benefits. As with the knowledge of Zhu Xi's formulation of "standard" domestic spatial practice, popular understanding of how the efficiency of this machine could be enhanced was disseminated in the late imperial period by consultations with technical experts (geomancers and carpenters), and by access to *users' guides*. These included illustrated excerpts from the *Carpenter's canon* included in the same household encyclopaedias and almanacs in which excerpts of Zhu Xi's work were prominently displayed. As print culture evolved, the market grew for specialist monographs on geomancy and for the services of geomantic experts.[38]

The material artefact of the altar thus incorporated two ideologies simultaneously. In the invisible architecture of decorum it embodied the hegemonic neo-Confucian ideal of social harmony and stability founded on filiality, virtue and collaboration. The altar was also the focal point of an energetic architecture of the house, which tied families, not into the ordered and socially harmonious space of the polity, but into an anarchic local landscape of competition between households; here it embodied an ideology of uncertainty and competition, in which each family must ruthlesslessly manipulate the local environment to its own advantage if it was to survive.

The shrine, then, was a key device in the normalization of the late imperial world order. The sociotechnical system of neo-Confucian values underpinned by domestic architecture was not static: it gradually expanded from a small new elite to incorporate almost the entire population of China, and it resisted dissociation for centuries despite the considerable social tensions generated by population growth, urbanization, commercialization, expansion into new territories and the reconfigurations of gendered and other divisions of labour discussed in later chapters. The altar celebrated the respectability of families who knew their place in society and strove for higher blessings than material success. At the same time people did want to succeed and opportunities were limited, they feared failure, and competition and uncertainties grew. The energetic functions of the domestic altar allowed families to take their fate into their own hands – and the blame was then theirs if things went wrong, for their technical skills had been insufficient. So the altar also acted as a safety valve, channelling potentially disruptive social energy in safe directions.

Technology as culture

This chapter has focused on the transformative effects of a technology that is conventionally considered "non-productive" in order to suggest the benefits of a

38 Feuchtwang (1974); He (1990); Smith and Kwok (1993).

more organic, anthropological approach to technologies and the work they perform. For the history of technology in non-Western societies such revisionism has clear attractions: it extracts the society in question from a system of negative comparisons with Europe to concentrate instead on material domains perceived as significant within that society. Moreover, by treating technologies *as* culture it constitutes an intellectual basis for the integration of history of technology into the broader historical discipline. The meritocratic elite of the Song justified their status through the social philosophy of neo-Confucianism and extended their influence and control through the elaboration of extended patrilineal descent groups. The domestic shrine was a material symbol of lineage ties and values. Starting in the Song dynasty the intellectual and political elites used the rituals and etiquette centred on the shrine, as well as rules for its material construction, to incorporate ever-widening circles of the population into the orthodox fold. Yet the shrine was not a simple instrument of orthodox discipline and social regimentation, for in a variant reading of ancestral powers it served as a device for transforming and directing cosmic energies. As a material artefact the shrine embodied ambiguities of meaning, and a corresponding moral flexibility, that aided its successful dissemination and made it a powerful instrument for the reproduction of the social order in the face of potentially disruptive forces.

2

INSTRUCTIVE AND NOURISHING LANDSCAPES

Natural resources, people and the state in late imperial China

> Does the presence of natural resources within the territorial confines of a state predispose its inhabitants to material prosperity and well-being or does it lead ineluctably to exploitation and immiseration? Are they a blessing bestowed by Nature or a curse made by Man? Even attempting to answer this riddle requires first trying to determine what exactly a natural resource is and coming to the conclusion that it does not so much exist as it is made to become.[1]

The scenes of rural life in the *Gengzhi tu* depict dense patchworks of tiny rice fields dotted with plantations of mulberry trees. Up to their knees in water, peasant farmers plough and harrow the mud to silken softness; bent double, they water, fertilize and transplant the delicate seedlings, weed them and weed them again (see Figures 0.1 and 6.2). In some scenes a daughter brings a welcome basket of food and tea; in others an official or a landlord watches the reapers or directs the filling of the granary. While the men toil in the fields the women toil indoors, washing and sorting silkworm eggs, caring for the voraciously munching worms as tenderly as delicate infants until the time comes to boil them alive in their cocoons. The women's labours do not end with the heaps of golden cocoons: next they must spin off the filaments, twist and reel the yarn and weave it into cloth (Figure 2.1).

Whether they were high-ranking officials or peasants choosing a New Year print, to the late imperial Chinese the busy fields and mulberry groves of the *Gengzhi tu* portrayed a landscape that was at once a material and a moral ideal: in a society

1 Bankoff and Boomgaard (2007b: 1).

FIGURE 2.1 Women weaving silk cloth on a drawloom, *Gengzhi tu*, 1742 imperial edition. A tall, spacious room was required to set up a drawloom, used for weaving figured silks. Here the loom is shown in a shed in the women's quarters of a well-to-do household. Both the weaver and the operator of the draw-threads are female.

where farming was the foundation not simply of the economy but of the social, moral and cosmic order of the imperial state, these scenes of intensively worked land and disciplined skills represented more than a source of material abundance. A landscape in which 'men ploughed and women wove', *nangeng nüzhi*, was also a setting in which humans acted out in microcosm the essential interflow of *yin* (female) and *yang* (male) energies that sustained cosmic harmony and social order. Throughout the late imperial era the Chinese state remained committed to this ideal landscape as both a goal and a tool of government, investing heavily in developing and maintaining versions of this technical-moral landscape across its territories.

The Chinese saw rice fields, or land that could be transformed into rice fields, as a blessing of nature, a resource made fruitful by human skill. As already mentioned in the Introduction, however, many modern historians have construed the type of landscape idealized in the *Gengzhi tu* as a historical trap. Fertile and carefully worked rice fields, it is argued, were the setting for a long process of technological and social involution, bringing over-population, deteriorating living standards and levels of environmental degradation and destruction far worse than anything seen in pre-industrial or even industrial Europe. Where the shaping of China's history is concerned, the riddle of the "resource curse" has provoked fierce debates about involution, growth with or without development, comparative levels of prosperity and of environmental damage, and the antecedents, dating and catalysts of the "Great Divergence".

This chapter traces the diffusion and adaptations of the iconic Chinese wet-rice landscape over time and through space. What I take as my theme, in tune with the goals outlined in the Introduction, is the long-term resilience of this landscape, its capacity to recover from crises and to sustain a series of what Jack Goldstone terms "efflorescences", notably in the Southern Song, later Ming and early to mid-Qing dynasties.[2] As with the ideals and practices of domestic space discussed in the previous chapter, I argue that the resilience and adaptability of the iconic rice farming landscape, both as an economic reality and as a political ideal, owed much to symbolic ambiguities as well as to material flexibility.

But first I would like to raise the question of how, when and why China's southern rice lands began to appear to the governing elite and to the population at large as a valuable resource, a source of plenty and of social order. Although nowadays it is commonly presumed that rice fields have fed China since time immemorial, here again – as with the new norms for domestic space – the rise of rice farming systems as the accepted model for good husbandry was a Song-dynasty phenomenon.

In the realm of Lord Millet

It is often thought that rice is and always has been the universal Chinese staple and its most typical crop. But that is not so. Until well into the Tang dynasty, the dry northern plains were the political and economic heartland of the Chinese state. The semi-legendary founding dynasties of Xia, Shang and Zhou all had their capital cities in the northern plains, where the sacred cereal, the source and staff of life, was not rice but setaria (or foxtail) millet. The mythical founding ancestor of the Zhou dynasty, the era to which all the classic rituals and key institutions of the Chinese state were traced back, was Lord Millet, *Hou Ji*, and it was Lord Millet who was honoured by later emperors at the annual spring rituals before the altar of the Gods of Soil and Grain, *sheji*.

2 Goldstone proposes "efflorescence" as the opposite of "crisis", 'a relatively sharp, often unexpected upturn in significant demographic and economic indices, usually accompanied by political expansion and institution building and cultural synthesis and consolidation' (Goldstone, 2002: 333).

Millets, with their myriad tiny grains, are well suited to dry regions such as north China. Common or broomcorn millet (*Panicum miliaceum*) and foxtail millet (*Setaria italica*) were the first cereals to be domesticated there, and by the first millennium BCE setaria was well established as the principal staple.[3] Other important northern grain crops included wheat, barley, and various legumes or beans, including soy.[4] Rice, however, was a minor crop in the north, restricted to naturally marshy areas and grown by rather primitive methods.[5] The other essential crops were textile crops: mulberry trees for feeding silkworms, or hemp, which was processed into the coarse but durable cloth that ordinary people wore.

Northern crop repertories and farming techniques reflected climatic conditions: rain was sparse and fell mostly in violent summer storms, winters were freezing and winds were harsh. The light loess soils characteristic of up-river regions were fertile but fragile, easily compacted or eroded. To coax plants into prospering, every effort had to be made to conserve any precious moisture in the soil, from winter snowfall to spring frosts. Farmers in low-lying areas or in regions downstream faced the opposite challenge of converting pasty alluvial mud into workable crumbs. An early lost work called *Hou Ji shu* (The book of Prince Millet) is quoted as follows in the *Lüshi chunqiu* (Spring and autumn Annals of Master Lü), a work on political and natural philosophy compiled in about 239 BCE:

> Are you able to make low, wet land fruitful? Are you able to conserve dry soils and temper them with damp? . . . Can you ensure that your millet heads will be rounded and the husks thin, that the grains will be numerous and plump so that food is plentiful? How may you do all this? By these

3 It was believed until quite recently that setaria was the earliest and most important domesticated staple of North China. New analytical techniques suggest that panicum, which is particularly resistant to drought, was in fact the earlier staple at least in Northwest China, while setaria came to predominate only from around 6000 BCE, as the climate shifted to a warmer and more humid regime; see Lu *et al.* (2009). Both types of millet were eaten steamed; glutinous varieties were used for brewing.

4 Even today millets, wheat and barley, sorghum and maize are the typical cereals grown and eaten in much of rural north China. Wheat and barley were introduced from Western Asia probably during the second millennium BCE (Flad *et al.*, 2010); wheat's popularity surged in Han times, when a range of hand and water mills came into common use and wheat could conveniently be ground into flour for consumption as breads or noodles (Servanti and Sabban, 2002: 4–8, 271–286). Sorghum reached China much later, from Africa by way of South Asia; maize was a post-Columbian introduction (Bray, 1984: 449–459).

5 Although early commentators usually glossed "the five grains", *wugu*, as including rice, *dao*, in addition to setaria, *ji*, panicum, *shu*, wheat and barley, *mai*, and legumes *shu*, some substituted hemp, *ma*, for rice (Bray, 1984: 432; hemp grain was not eaten but pressed for oil). Glutinous rice, used for brewing, seems to have been grown as commonly as ordinary rice in the north, and dry rice as commonly as wet rice. The sixth-century farming treatise *Qimin yaoshu* names 11 glutinous and 12 non-glutinous rices (*QMYS*: 99–100); wet and dry cultivation techniques are both treated in just a few paragraphs (*QMYS*: 99–107).

> fundamental principles of tillage: The strong [soils] must be weakened, the weak strengthened. The rested must be set to work, the hard-worked rested. The lean must be fattened, the fatter made leaner. The compact must be loosened, the loose compacted. The damp must be dried out, and the dry dampened . . . Till five times and use the weeding-hoe five times. Observe all these rules closely.[6]

The classic treatise on northern farming practices, the *Qimin yaoshu* (Essential techniques for the common people), completed c. 535 CE by a Northern Wei official named Jia Sixie, quotes this excellent if somewhat inscrutable advice near the beginning of the first chapter, illustrating it with a quotation from a Han dynasty work, the *Fan Shengzhi shu* (Book of Fan Shengzhi):

> When the apricots start to bloom, plough light soils and weak soils immediately; plough them again when the apricot blossoms fall. After ploughing roll down the surface at once . . . If the soil is very light it should be trampled by cattle or sheep. This is what is meant by 'making weak soils stronger'.[7]

By the time that Jia Sixie was writing, animal-drawn rollers and several different types of harrow were used to break up, smooth and compact fine soils, and operations were carefully timed to avoid damaging soil structure or losing moisture. Commenting on the *Qimin yaoshu* in his *Nongshu* (Agricultural treatise) of 1313, the Yuan official Wang Zhen noted that in his day, in order to prevent the soil drying out, northern farmers ploughed early or late in the day during spring, and in summer they ploughed at night; only in autumn, when the soil was soaked after the rains, did they plough at midday.[8]

The technique of planting in rows was also favoured from very early times:

> If crops are grown in rows they will mature rapidly because they will not interfere with each other's growth. The horizontal rows must be well drawn, the vertical rows made with skill, for if the lines are straight the wind will pass gently through.[9]

6 *LSCQ:* 27–28.

7 *QMYS*: 27. Fan Shengzhi is said to have been an agricultural official in the capital district. The *Fan Shengzhi shu*, dated c. 10 BCE, was one of nine works by "agriculturalists" (*nongjia*) listed in the bibliography of the official history of the Western Han dynasty. Most were recorded as being of considerable length; all are now lost except for quotations in later works (Wang Yuhu, 1979: 6–17). On the *Fan Shengzhi shu*, see Bray (1984: *passim*) and Zeng (2008: 230–245). On the *Qimin yaoshu*, see Bray (1984: *passim*), Zeng (2008: 260–328) and Bray (forthcoming).

8 *WZNS*: 23.

9 *LSCQ*: 67.

The benefits of row-planting were further increased by ridging, which kept the built-up soil moist around the roots of the plants and concentrated the benefits of fertilizers. A field could be ridged using hoes, but turn-ploughs were more effective. Ploughs with cast-iron mouldboards for turning the furrow were already in wide use by Han times, as were iron-shared seed-drills (some with a single drill, others with two or three) that allowed seeds to be sown with regular spacing along the ridges. Row-planting and ridging also made hoeing easier once the seed had sprouted; repeated hoeing was yet another way to keep water from evaporating from the soil surface, and to cut down competition for precious moisture from weeds:

> You should not mind how many times you hoe setaria; once you have been right round the field start again, and do not stop even for a short time simply because there are no weeds. (Hoeing does not just get rid of weeds, it keeps the soil at a good tilth (*shu*) and will give full ears, with thin husks, that do not shatter. If you hoe your field ten times you will get 'eight tenths grain'.)[10]

In many fertile regions of the north, it seems that even by Han times populations were too dense for land to be allowed to rest between crops. Instead of fallowing their fields, as was common in Europe well into the early-modern period, these Chinese farmers supplemented animal manure, straw or turned-in stubble with green manures such as alfalfa or beans, planted early and ploughed in before nitrogen-hungry crops such as cereals were sown. Crop rotations were also well developed. The *Qimin yaoshu* indicates the best preceding crop, *di*, for all the main cereal crops as well as for important oil and fibre crops, and even for some vegetables or herbs. Thus a barley crop might be followed by rape turnips (grown for their greens but also for oil), followed by setaria, then adzuki beans, then hemp or squash.[11] Even so, it seems that yields for setaria were unlikely to surpass one tonne per hectare.[12]

The sixth-century *Qimin yaoshu* lists nearly 100 varieties of setaria millet, bred for different ecologies or uses, for their colour, flavour or culinary qualities, or to resist drought, flooding or insects; the chapter on methods for cultivating setaria is longer than any other in the book.[13] Right through the Tang and Song dynasties setaria remained the principle crop of the north and a significant component of the tax system. The eighth-century Hanjia tax-granary at the Tang capital of Luoyang (sited on a tributary of the Yellow River, in modern Henan province)

10 That is, the weight of the hulled grain will be eight tenths that of the grain in the ear; *QMYS*: 44.
11 Bray (1984: 431).
12 Bray (1984: 448).
13 Bray (1984: 434–448); *QMYS*: 42–72.

was built exclusively for storing millet, which, unlike most other cereals, kept best underground. With a total capacity of 6 million bushels, *shi*,[14] the Hanjia granary consisted of nearly 300 huge cup-shaped pits, ranging in size from 8 m across by 6 m deep to 18 m across by 12 m deep. The pits were lined with smooth clay, fired to make them waterproof and lined again with wooden boards and rush matting; the opening was likewise covered with wooden slats and matting, then sealed with plaster. Thus kept cool and dry, the millet would keep for several years, and the stocks could be used for price regulation or famine relief as well as for the more routine purposes of feeding troops and paying government officers.[15]

Villagers also routinely used underground pits to store their millet, albeit on a more modest scale: not only did the grain keep well, but – as the agronomist and official Wang Zhen noted in his treatise of 1313 – pits concealed the grain stores from the eyes of robbers or marauding troops: if one sealed the mouth of the pit with a clod of turf, then a stranger would never notice it; better still, one could plant a small tree over the pit, which would not only conceal it but also indicate the state of the grain underneath, for should the grain start to rot the leaves would wither and turn yellow.[16]

Despite the increasing importance of the Southern economy, it was the millet-centred landscape of North China that furnished Chinese norms of skilful farming, and the iconography of rural productivity, from Han times well into the Song dynasty.[17] The main crops were dry-land cereals such as millets. Successful husbandry required the use of a range of animal-drawn equipment including plough, seed-drill, harrows and rollers, as well as hand-tools such as hoes; the animals and labour to keep the fields in good heart; and sufficient land to rotate crops. In other words, this was a technical system that rewarded economies of scale, working better for large, well-resourced estate farms than for small peasant holdings.[18]

The interests of the state, however, lay in supporting a quite different style of land management. Its revenues were predicated upon a tax base of small peasants who paid land tax in kind – usually assessed at about a tenth of the grain and cloth they produced each year. While peasant farmers found it difficult to avoid taxes, powerful landowners had many ways to resist full payment. During the four centuries of the Han dynasty (202 BCE – 220 CE), officials fought constantly to protect their smallholder base against the depredations facilitated by a free market in land. Among the techniques they used to protect and develop peasant farming were loans, tax

14 A *shi* was approximately equivalent to 60 kg in weight.

15 On the Hanjia granary, excavated in the 1970s, see Bray (1984: 401, 416).

16 *WZNS*: 294–295.

17 See Chapter 8. Typical early images of northern farming scenes and landscapes are reproduced in Bray (1984) and Wang Chaosheng (1995).

18 This discussion of the political tensions inherent in developing northern farming is based on Bray (1984: 587–597).

relief in bad years, and the distribution of cheap cast-iron implements mass-produced in state-controlled foundries.[19] Despite these efforts, peasant-scale farming in the northern heartlands remained vulnerable and underproductive – not least because small farmers could not afford the best practices, the animals, the equipment and the intensive methods just described. Large estates expanded steadily as wealthy families bought land from failed peasants, who often ended up working as their labourers. In times of unrest, estate owners could also offer protection, and many peasants voluntarily gave up their land in exchange for security.[20]

Han officials lamented and legislated in vain as peasant-held lands dwindled, estates expanded and state revenues declined. Often the state attempted reforms, invoking the principles of egalitarian land distribution supposedly in force during the golden days of the early Zhou dynasty, where peasant families were given plots of land of equal size grouped around a communal well. The "well-fields", *jingtian*, were the inspiration for the "equal fields" system, *juntian*, introduced by the first emperor of the Northern Wei dynasty in 485 CE and retained by succeeding governments through the Sui and early Tang dynasties until the mid-eighth century.[21] Such legislation never succeeded, however, in eliminating the large estates or stemming the continuous drift of peasant lands into landowners' control. Nor could peasant farms match estates in quality of techniques or in output. From the point of view of agronomists of the period, it was large estates that constituted the models for best practice.[22]

Jia Sixie, the sixth-century author of the *Qimin yaoshu*, served as a prefect under the Northern Wei, which had introduced the system of equal allotments.[23] Despite his status as a state official, and despite entitling his treatise 'Essential techniques for *the common people*', Jia's treatise advocates methods of farm management that lay far beyond the scope of a peasant farmer. In addition to the basic tasks of food and fibre production, Jia discusses commercial activities such as planting timber,

19 On the prolonged and fierce debates concerning whether or not the Han state should have a monopoly of iron production, see Wagner (2008: 171–245).

20 Hsu (1980).

21 The *juntian* allocation under the Northern Wei (386–535) was 140 *mu* (approximately 8 hectares) for the household of a married man between the ages of 21 and 59: 100 *mu* for grain and 40 for mulberries or hemp. During the Tang dynasty (618–906) the total was reduced to 80 *mu*, of which one quarter was permanently owned, and the upper limit for aristocrats' estates was set at 100 *qing* (1 *qing* = 100 *mu*). On the challenges of implementing this system, and its historical vicissitudes, see, for example, Wright (1979) and Tang (1990).

22 This was equally true during times when markets flourished and in times of unrest when self-sufficiency was the goal. See Ebrey (1974) on the farming system advocated by Cui Shi, a member of a powerful landowning family, in his *Simin yueling* (Monthly ordinances for the four classes of people), composed around 160 CE. See Zeng (2008: 325–328) on how Jia Sixie, the author of the *Qimin yaoshu*, approached commercial production.

23 Bray (1984: 55–59); Bray (forthcoming).

raising sheep, growing oilseed or brewing, all on a generous scale. A recipe for the brewing of "spring wine" calls for 180 bushels of grain, the recipe for the popular salty condiment called *chi* for a thousand bushels.[24] The chapter on oilseed says:

> 1 *qing* [100 *mu*, approx. 6 hectares] of rape turnip (*wujing*) will yield 200 bushels of seed. If you take the seed to an oil-pressing establishment they will bring you three times their weight in hulled grain, that is 600 bushels. This is a far better yield that you would get from 10 *qing* of grain.[25]

In other words, Jia Sixie's unit of calculation for growing rapeseed equalled the total official land allocation for a peasant family. This illustrates a basic contradiction at the heart of the northern dry-land farming system. From the statecraft perspective, the desirable landscape was morally and politically as well as materially defined as a landscape of myriad peasant smallholdings growing millet and hemp. From the perspective of the agronomist or of anyone belonging to the landed elite, the desirable landscape was a large, well-managed and profitable estate, well equipped with labour, draft animals and equipment.[26]

Lands of fish and rice

Although it was a rarity in the north, wet rice[27] had been the staple crop in the Yangzi region and in southern China for millennia.[28] The powerful southern states that vied with northern rivals for supremacy until the unification of the empire in 221 BCE – including Chu and Wu along the middle and lower Yangzi (later known as Jiangnan), Yue on the eastern coast, the kingdoms of the Canton region (Lingnan) and Shu in what is now Sichuan – all depended on rice for their wealth and power. So too did the succession of southern dynasties that governed from the capital of Jiankang (modern Nanjing) following the fall of the Han and the break-up of the empire after 221 CE.

Polders, irrigation tanks, bunded fields and probably transplanting had all been developed within these regions very early.[29] Although the most sophisticated rice-farming techniques were probably still confined to just a few areas, even at the end

24 Sabban (1993: 98, 92).
25 QMYS: 133.
26 See Bray (1984: 587–597).
27 Rice grown in standing water, in fields that must be levelled and surrounded by a low mud dyke or *bund* to keep in the water. Dry rice, which depends on rainfall alone, can be planted on any terrain; around tropical Asia it is often grown by shifting cultivators on steep, recently cleared hillsides (Bray, 1984: 496–498; Grist, 1975: 179–196; Scott, 2009).
28 Zhang and Hung (2010).
29 Bray (1986: 77–90).

of the Han the southern wet-rice landscapes were sufficiently productive to accommodate not only the needs of local populations and governments, but also huge influxes of northern refugees. Peasants who had fled south must presumably have learned to grow rice instead of millets, despite their notorious reluctance to acknowledge the new landscapes as home.[30] Yet, knowing nothing themselves of the work involved in intensive rice farming, and all too familiar with the hard labour needed to wrest a good crop of millet from the soil, northerners still referred enviously to the southern regions as "lands of fish and rice", *yumi zhi xiang*, lands of natural bounty where it seemed that food could be grown and gathered without effort.

By the early Tang dynasty (618–906), both the southern population and the acreage under cultivation were expanding rapidly and the south was becoming more solidly incorporated into the reunited imperial polity. Migrants from the crowded northern plains continued to stream south in search of land, and now tax-rice also flowed north to the capital along recently built or improved north-south axes of communication, including the Grand Canal. By the mid-Tang, southern taxes in rice were beginning to match northern taxes in millet.[31] The acreage and the productivity of southern wet-rice farming were increasing steadily. This was due in part to official investment in infrastructure. Li Bozhong lists dozens of major hydraulic projects constructed in Jiangnan during Tang times, including dams, polders and irrigation networks on inland rivers, and the construction of sea walls to reclaim coastal lands.[32] At a more modest level, small groups of farmers could cooperate on small-scale techniques for raising, impounding, draining and channelling water – including small-scale irrigation tanks, aqueducts constructed from bamboo tubes to carry water across valleys (Figure 2.2), and chain-pumps to raise water.

Although the documentary evidence is circumstantial rather than unequivocal, Li Bozhong argues that the characteristic toolkit of intensive rice farming so lovingly depicted in the twelfth-century *Gengzhi tu* – the iron-shared turn-plough, the iron-tined harrows and ribbed rollers used to puddle the clods of the paddy fields into a smooth, level, silky mud, the repertory of hoes for tilling or weeding, and even the pedal chain-pump (see Figure 0.1) – were coming into common use in Jiangnan during the Tang, aided by the development of the southern iron industry.[33]

30 Some historians believe that during the period 290–313 alone, two million refugees fled south from the Central Plains to avoid nomadic invaders (Crowell, 1990: 175). Crowell describes successive southern governments' long and fruitless battles to register (and tax) these reluctant settlers, who always claimed that their homes were still in the north.

31 Twitchett (1963); Li Bozhong (1990: 285–293).

32 Li Bozhong (1990: 77–82, Table 3.1).

33 Li Bozhong (1990: 73–98). It is interesting that the first detailed technical description of the turn-plough comes from the south. The *Leisi jing* (Classic of the plough, tr. Bray, 1984: 181–183), was composed c. 880 by Lu Guimeng, a recluse who lived near modern Suzhou in Jiangnan.

FIGURE 2.2 Bamboo aqueduct, 1530 edition of Wang Zhen's *Nongshu*.

Not all southern wet-rice farming was intensive during the Tang. In many areas the rice was not sown in nursery beds and transplanted, but sown directly into the main field; often paddy fields were not irrigated but depended on natural rainfall; and it was still common to leave rice fields fallow one year in two and use fire to clear the weeds. (In fact extensive farming persisted throughout the imperial period, for example in newly opened frontier zones, or where labour was short.) But historians such as Li Bozhong argue that during the Tang, intensive rice-farming techniques became routine in key areas of Jiangnan and Lingnan, setting norms for the improvement of rice cultivation elsewhere. The material contours of China's characteristic wet-rice landscapes, continuously refined, developed and disseminated in later dynasties, were already clearly traced in Tang-dynasty Jiangnan.

The field-types of wet-rice farming, like the toolkit, were distinctive. In wet-rice farming, where precise control of water depth and flow was crucial, small fields were easier to manage and more productive than large fields. But the small fields were characteristically clustered together as subdivisions of larger irrigation units (Figure 2.3). Water flowed across and between adjacent fields when the farmer breached the bund with a hoe: this apparently primitive technique served to share out water from the common source, and to adjust water levels or temperature to the needs of the growing plants.[34]

34 On the contradictions between common management of water and individual management of fields or ownership of land, see Bray (1986: 66–68, *passim*), Lansing (2006) and Li (2010).

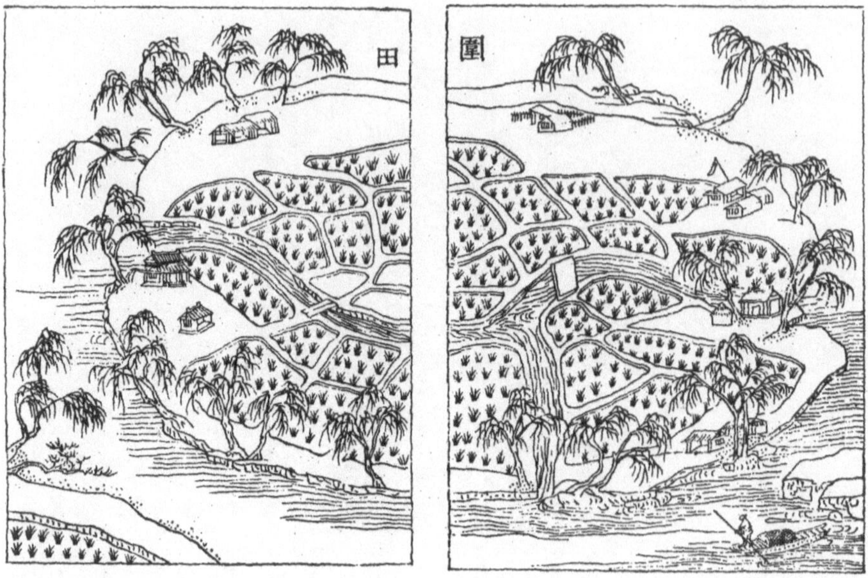

FIGURE 2.3 Poldered or "encircled" field, *weitian*; this illustration from the imperially commissioned compilation *Shoushi tongkao* of 1742 is based on an original in Wang Zhen's *Nongshu* of 1313.

Improvements in water-control technology allowed new types of land to be converted to paddy fields and facilitated more intensive cropping patterns. If fields were thoroughly drained after the rice harvest, for instance, a dry crop such as wheat or barley could be grown over the winter. If the rice was sown in small nursery beds before transplanting into the main field, there might be time for two or even three crops in the main field.

Higher-lying fields required irrigating; low-lying fields were as likely to need draining. A hamlet or even a single household could easily construct a set of small tanks or ponds for gravity irrigation:

> On high land identify the places where water accumulates and dig out ponds. Out of 10 *mu* of land [approx. 0.6 ha] you must be prepared to set aside 2 or 3 *mu* for water storage. At the end of the spring when the rainy season comes, heighten the banks and deepen and widen the interior. Strengthen the banks by planting mulberry or *zhe*[35] trees, to which buffalo may be tethered in the shade as their nature requires. By trampling the banks the

35 The silkworm thorn, *Cudrania tricuspidata*, a close relative of the mulberry whose leaves are also used to feed silkworms.

> buffalo will strengthen them, the mulberries will be well watered and grow into fine trees, and even in the dry season there will be sufficient water for irrigation, yet in heavy rains the tank will not overflow and harm the crops.[36]

This passage, from the *Nongshu* (Agricultural treatise) of 1149 by Chen Fu, illustrates the integration of rice farming and sericulture typical of much of Jiangnan. As alternatives to tanks, farmers with fields on higher land might divert or build channels off a stream, or construct bamboo aqueducts to bring water from another hillside.[37]

As pumping technology improved, low-lying lands were increasingly brought under cultivation. *Weitian* ("encircled" fields) and *yutian* ("poldered" fields) were usually constructed either by powerful families or by the state: tracts of low or marshy land along lake- or riverbanks were surrounded by a dyke and divided up into small bunded fields, separated by drainage channels. Ranging in size from a few hectares to several square kilometres, these fields were highly valued because they were secure from flood and drought, and fertile enough to produce large, marketable surpluses.[38] *Guitian* ("strongbox" fields) were a smaller version:

> In strongbox fields the soil is built up [into a high dyke] to protect the fields . . . On all four sides are placed flumes. The shape is similar to a strongbox and convenient to cultivate. If the area is very marshy then the fields should be of smaller size, built with firm dykes on higher ground so that the water outside cannot easily get in but the water inside can conveniently be pumped out with a chain-pump, *che*. The parts that are shallowly flooded should be sown with yellow quick-ripening rice, *huangludao*, which only takes 60 days from sowing to harvest, and thus is not endangered by floods. Where the water is excessive and aquatic plants start growing of their own accord, then barnyard millet, *xianbai*, can be harvested.[39] In the high, well-drained areas all types of dryland crops can also be grown. These crops will all stave off famine. This is the best way of reclaiming marshy land.[40]

36 *CFNS*: 2.

37 Bray (1986: 69–90).

38 *WZNS*: 186–187; Bray (1984: 114). In her interpretation of Lou Shou's poem and pictorial depiction of a seemingly peaceful scene of a chain-pump at work irrigating fields in a *weitian*, Hammers notes that when it came to reclaiming land, one community's gain was often another's loss. The apparently universal advantages of technological innovation, then as now, often proved to benefit the rich and strong rather than the poor (Hammers, 2005: 135).

39 Barnyard millet (*Echinocloa crus-galli*) is a small-grain cereal that thrives under similar conditions to rice. It is sometimes considered a crop, sometimes a weed, sometimes a famine food; see also below on the reclamation of salt-lands.

40 *WZNS*: 188–189; Bray (1984: 114–115). Zeng (1998) argues that *huanglu* quick-ripening rices, which grew well in swampy conditions, played an even more critical role in the expansion of rice production post-Song than the equally quick-ripening Champa rices (see below). Their advantages were complementary, since Champa rices were suited to drier fields including terraces.

This passage, from the *Nongshu* (Agricultural treatise) of 1313 by Wang Zhen, describes a suitable strategy for underprivileged peasants who could only gain access to marginal and highly vulnerable lands, where they constantly risked poor harvests or famine. Poldered and "encircled" fields, however, were a source of wealth, much coveted by powerful families and their tenants – but when too much of a lake was drained, or a river's flow was impeded, disastrous flooding would ensue. Throughout late imperial times officials struggled to keep these forms of reclamation (and of gentry empowerment) under control, while recognizing that new land was needed to feed the growing population.

Even where powerful families owned huge tracts of rice land, however, rather than operating large-scale farms they found it advantageous to rent out the land in small farms to skilled peasant tenants. As one late-seventeenth-century landlord put it:

> There is a proverb which says "It is better to have good tenants than good land" . . . There are three advantages in having good tenants, namely that they are on time with ploughing and sowing, they are energetic in fertilizing, and they are resourceful in conserving every drop of water.[41]

A Jiangnan landowner writing a few decades earlier advised keeping only 10 *mu* (approx. 0.6 ha) for the home farm and renting out the rest.[42]

Less environmentally and socially contentious than the huge polders of the inland lakes and rivers was the reclamation of salt-lands along the coast. In 1026 the Song official Fan Chengda organized the construction of several hundred *li*[43] of sea wall, *haiyan*, along the coast of what is now Jiangsu province. Reclamation took place in several steps:

> Along the seashore farmers practise a system of stepped fields, *dengtian*. The mud and silt brought in by the tide pile up to form little islands which are sometimes sucked away by the whirling currents, so that the exact area of land is difficult to calculate. On top [of the silt] grow thick clumps of salt grasses. The farmers wait until there is a high tide, and then they gradually tickle the mud into life. First they sow barnyard millet, and when this has eradicated the salinity of the soil the land can be used for crops. 'Wash away the salt so rice and millet grow', says the poem.
>
> Where river banks overflow, and along the sea shore, walls are build or standing posts are planted to prevent the tide encroaching. At the side of the

41 Zhang Ying, *Hengchan suoyan* (Remarks on real estate), probably completed 1697, tr. Beattie (1979: 149); see also Bray (1984: 82–85).

42 *BNS*: 115.

43 During the Song dynasty 1 *li* was approximately 300 m.

> fields drains are dug to draw off surplus rainwater and to irrigate the field in times of drought; these are called sweet-water drains, *tianshui gou*.
>
> The yields of silt fields, *tutian*, may be as much as ten times those of normal fields, and the common people often take them as heritable property, *yongye*.[44]

No wonder that hundreds of little peasant boys were named after Fan Chengda in the years following the successful completion of his sea wall project.[45]

Finally, the list of ingenious field types would not be complete without mention of terraced or "ladder" fields, *titian*. Although nowadays most people think of rice terraces as the archetypal Chinese landscape, in fact they seem to have originated in the mountainous inland regions inhabited by Dai, Miao and other non-Chinese populations. Han Chinese officials despatched to exotic southern locations began remarking upon terraces during Tang times. By the fourteenth century, Wang Zhen felt sufficiently confident to provide the first proper technical description in Chinese:

> Terracing means cutting steps in the mountain to make fields. In mountainous areas where there are few level places, apart from stretches of rock, precipices or similarly barren areas, all the rest, wherever there is soil, from the valley bottom right up to the dizzying peaks, can be split to make ledges, *deng*, where crops can be grown. If stones and soil are in equal proportion then you must pile up the stones in rows, encircling the soil to form a field. There are also mountains where the slope is extremely steep, without even a foothold, at the very limits of cultivation where men creep upwards bent close to the ground. There they pile up the soil like ants and prepare the ground for sowing with hoes [because the fields are too narrow to use ploughs], stepping carefully while they weed for fear of the chasm at their side. Such fields are not stepped but mount like the rungs of a ladder, hence their name of 'ladder fields'.
>
> If there is a source of water above the field, then all kinds of rice may be grown, but if it is only suitable for dry crops then setaria millet and barley are best.[46]

Nascent during the Tang, most of the technical features of intensive wet-rice farming were widespread through Jiangnan and other southern regions by the early Song dynasty. Yet although southern wet-rice farmers now provided the bulk of the state's grain revenues and the empire's wealth, until the fall of the Northern Song, China's political and social elites continued to draw their ideals of agronomic

44 *WZNS*: 192.
45 Bray (1984: 121).
46 *WZNS*: 190–191; Bray (1984: 123–126).

expertise and the iconography of politically desirable and technologically sound farming from northern repertories of dry-land farming. Although government policies for agricultural development during the Northern Song targeted south and north equally, there were still no specialist treatises on southern farming systems to rival the great treatises on northern methods.[47] All this was to change radically in the few decades following the loss of the northern provinces in 1126. The rice- and silk-farming landscapes of Jiangnan rapidly achieved iconic status, and agronomic treatises shifted to a focus on southern cropping systems. This transformation was durable, I argue, largely because the technical dynamics of irrigated rice farming meshed closely with the political goals of late imperial China.

The mandate of the agrarian state

The ideological goals of imperial statecraft were to preserve and extend the political and moral order throughout its territories. A primary concern was the welfare of the people: the task of the ruling class was to guarantee peace and prosperity to their subjects, partly by ensuring them a stable livelihood, and partly by instilling them with ethical values. Floods or famines were signs that a ruler had failed to fulfil the mandate of heaven and justified a rebellion to overthrow him; a local magistrate was responsible for fighting poverty as well as collecting taxes; and officials who brought dramatic improvements to the people under their charge, for instance by constructing a dyke to keep flood tides at bay, would be rewarded by having hundreds of peasant babies named after them. To this moral responsibility of the ruler was joined the more pragmatic concern that strong states depended upon a healthy economy. A well-nourished peasantry and full state granaries were needed to support the expenses of government; to maintain large and well-equipped armies to defend the frontiers and extend the empire's territories; and to carry out public works.

The core of the Chinese fiscal system throughout the imperial era was the land tax. Over time the ways in which tax dues from peasant farmers and landowners were calculated varied, as did the rates of taxation.[48] Except in certain times of extreme crisis, however, rates were usually kept low, equivalent to around one-tenth of the harvest.

> Government taxation rates were set by officials who believed that light taxation allowed the people to prosper, and since a prosperous people was

47 See Wang Yuhu (1979) and Hu (1985). The first official printing of agricultural books took place in the early eleventh century, when two treatises were printed by the Song government for circulation to magistrates. One was the *Qimin yaoshu*, the other was the *Sishi zuanyao* (Compendium of essentials for the four seasons), a ninth-century farming calendar. Both works described northern farming (Hu, 1985: 46).

48 For example, Huang (1974); Wong (1997: 131 ff); Mazumdar (1998: 211 ff, 288 ff).

> held to be crucial for the maintenance of a powerful state, tax rates were low. Some officials, at least, no doubt realized that attempting to raise tax rates would have the unhappy consequences of promoting resistance and even rebellion.[49]

The Chinese ideology of empire envisaged a social contract based on reciprocity between the emperor and his millions of peasant subjects. The emperor, through his officials, offered protection and benefits to farming households, who in return worked diligently to pay the taxes without which the empire would founder. This basic social contract made rural production the lifeblood of empire. Peasant men and women were acknowledged as direct participants in the reproduction of the state, and throughout imperial times the state repeatedly intervened to reduce the power of any group (feudal aristocrats, local strongmen, Buddhist monasteries or wealthy landlords) who threatened to siphon off excessive amounts of produce or land, take peasant families out of the direct tax system, or accumulate sufficient assets to resist or conceal all or part of their tax dues. The goal was to maintain a direct fiscal relation between peasant producers and the state, to keep the number of those peasant producers as high as possible, and to raise levels of peasant production of the basic commodities of grain and textiles wherever feasible.[50]

The goals of empire shaped the techniques of government and the historical evolution of the relations between production, consumption and government, as R. Bin Wong argues in his discussion of the specificities of what he calls the Chinese *political economy*,[51] or what we might alternatively refer to as a *mode of production*. If it was both the moral responsibility, and the most effective method of government, to bring benefits to the people, *limin*, by educating and nourishing them, *jiaoyang*, then clearly government had to play an active role not only in protecting the peasantry from other groups anxious to extract resources from them, but also in providing the resources and knowledge that would enable them to prosper.

Imperial China was from its inception an agrarian state in the strong sense of the term. The promotion of agriculture, *quannong*, was at the core of the philosophy and techniques of government. Even before the first unification of the empire in 221 BCE, Chinese states were actively involved in producing and circulating farming

49 Wong (1997: 90).

50 Commerce, by contrast, was a very minor source of state income. After the end of the Song (1279) the taxes on most forms of internal trade were minor or non-existent. The late imperial state was able to spare merchants from taxation in part because the fiscal mechanisms it had devised tapped agricultural revenues quite effectively (Wong, 1997: 133).

51 Wong (2002).

knowledge, building infrastructure and managing distribution.[52] This interventionist approach continued throughout the imperial period and involved every level of government, from the emperor and his senior ministers down to local magistrates.

I have already referred to a range of *quannong* issues and policies undertaken during the earlier dynasties to promote and protect peasant farming in north China. The Western Han state (206 BCE – 8 CE), for instance, offered loans and tax relief, and organized the production and distribution of iron implements, while in 485 CE the Northern Wei rulers established a system of peasant land allotments that remained in place for several centuries. Yet after each reform large estates began once more to displace peasant smallholdings. I have argued that one reason why rulers were unable to maintain a robust, self-sustaining peasant sector and to curb the power of landowning families in north China was that, under the conditions of northern dry-land farming, large-scale operations were much more productive, profitable and technically sophisticated than smallholdings.

The dynamic of technical and social development was quite different, however, in the south where wet rice was the main crop. If we survey the history of wet-rice farming in pre-modern Asian societies, we see clearly that once intensive methods were adopted, small farms consistently performed just as well, and adopted new technologies just as successfully, as large ones.[53] Wet-rice farming constitutes what I call a *skill-oriented* technical system, responding favourably to low-capital inputs and the intensive use of human skills.[54] The following sections show why the economies of scale typical of dry-land northern farming simply did not operate in the southern rice regions: on the contrary, as techniques improved, cropping patterns intensified and yields per acre rose, smaller units of management routinely outperformed large. Intensive wet-rice farming spurred the growth of highly diversified and commercialized rural economies. This growth certainly contributed to the expansion and prosperity of the landowning and merchant classes. But because smallholders were recognized as the most efficient farmers, gentry landowners rented out all but a few acres to peasant smallholders. Since rent-payers, like independent peasants, were all recorded on the tax registers, the landscape matched the state's ideal of a realm made fruitful and prosperous by a multitude of tax-paying peasant subjects. Whether or not this historical process was *involutionary* has been a matter of fierce debate, which I take up later. But let me first return to the Song dynasty and the rapid ascent of Jiangnan rice farming as an icon of productivity.

52 See Chapter 6 on *quannong* and innovative knowledge.

53 Bray (1986).

54 Bray (1986: 113–116). The logic, I argue, does not apply to pre-modern systems alone. See Francks (2014, in press) for an analysis of how skill-oriented rice farming and its sociotechnical legacy has influenced the industrialization and modernization of the Japanese economy.

Nourishing and instructive landscapes

The Song dynasty (968–1279) marked the beginning of a new era in Chinese history. The old aristocratic families who had hitherto formed the ruling elite were replaced by a meritocracy ranked by a national examination system; printing of books, records and official documents became routine; cities grew, markets for consumer goods expanded and commercial networks began to penetrate the countryside.[55] Although Song political elites argued just as strenuously over the principles of state economic intervention as their Han predecessors, the *quannong* policies of the period included costly investments in hydraulic projects, the dissemination of technical information and new crop varieties, and cheap loans to poor farmers. Many state-sponsored improvements in farming took place in the Northern Song, notably in the context of Wang Anshi's programme of New Economic Policies, *xinfa*, introduced between 1068 and 1085 to aid the poor and increase state revenues by taxing the rich and setting up state monopolies in industries such as tea.[56] But the most enduring transformations were achieved under the Southern Song.

The loss of the northern provinces in 1126 precipitated a crisis that permanently transformed perceptions of the relative importance of the northern and southern rural economies. The government, the propertied elite, and huge numbers of northern peasants fled south, and suddenly the southern territories were no longer remote provinces – they had become the heartlands of the Chinese state. Not only were the southern rice landscapes coaxed into accommodating the newcomers and yielding food for millions of extra mouths, their abundance helped build the most magnificent capital city that China had ever seen (in what is now Hangzhou, known to Marco Polo as Kinsay), and their productivity powered several centuries of economic growth.

The loss of the north was hardly unforeseen, and well before 1126 Song administrators had been encouraging the development and dissemination of improved rice technology in the southern provinces. One of the most famous measures was the introduction to the Yangzi Delta in 1012 of new varieties of quick-ripening rices from Champa in Vietnam.[57] This transformed production

55 Kuhn (1987, 2009).

56 Historians differ as to how much these policies in fact aided the poor; see Golas (1980) and Smith (1991).

57 Bray (1984: 492–495). The "Champa", *zhan*, rices (already cultivated in Fujian before their introduction to Jiangnan) were long-grained *indica* rices, *xiandao*, whereas the typical local rices (including the quick-ripening *huanglu* rices mentioned earlier in the context of land reclamation) were round-grained *japonicas*, *jingdao* or *gengdao*. Initially most Jiangnan people still preferred to eat japonicas whenever possible. Although the highly adaptable *xian* soon became popular with farmers through central and southern China, local tastes remained highly specific and many eighteenth- and even twentieth-century consumers retained a marked preference for *jing* rice (Cheung, 2014, in press; Lee, 2014, in press).

patterns, allowing two crops of grain (a summer crop of rice and a winter crop of wheat or barley) to be grown in a year. Seeds of the new varieties were distributed to farmers through the district *yamen*, and written instructions on their cultivation methods were circulated to magistrates and to literate "master farmers", *nongshi*, chosen for their skill and experience to fill a minor official post that carried the duty of improving farming techniques in their village. They were to instruct their peers not only in new techniques such as improved sowing and fertilizing methods or crop choices, but also in the organization of mutual aid and so on. Agricultural treatises were printed by the Song government and circulated, and the state also provided infrastructural support such as new irrigation networks, as well as low-interest loans, lower taxation and tax rebates for farmers who opened up new rice-land.[58] The effects of these policies were so dramatic that the historian Mark Elvin has referred to them as "the Song Green Revolution".[59]

It was in the Southern Song period that the virtues and skills of rice farming first became recognized icons of a quintessentially Chinese social order. The Song government-in-exile took the local Jiangnan landscape to its bosom, and gave rice a prominent symbolic role in official discourse. The new political cult of the rice economy was crystallized in the history of the *Gengzhi tu*, a pair of scrolls depicting local processes of rice farming and silk production, painted during the 1130s by Lou Shou, the magistrate of Yuqian County just a few miles from Hangzhou. The *Gengzhi tu* was enthusiastically received by the emperor just a few years later, and soon it was famous around the country. The story of the long imperial love affair with the *Gengzhi tu*, and of the prominence that the work rapidly assumed in official and popular visual culture as well as in *nongquan* campaigns over the centuries, is told in detail in Chapter 8.

The *Gengzhi tu* was the first work to celebrate the indispensable role of the irrigated Jiangnan landscape in supporting the imperial order, and to show the skills, hard work and technical sophistication that made this contribution possible. But it is no coincidence that only a few years later, in 1149, an elderly gentleman-farmer named Chen Fu completed a short but richly detailed agricultural treatise, *nongshu*, that systematically described the rice- and silk-farming system of Jiangnan, not in pictures and verse, but in the sober prose style that had been developed several centuries earlier for technical description.[60]

Chen Fu's *Nongshu* is divided into three sections, the first on cropping (mainly of wet rice), the second on caring for water buffalo (the indispensable draught animal

58 These policies and their application are described in detail in Golas (1980).

59 Elvin (1973).

60 See *CFNS*; Miao Qiyu (1981a); Wang Yuhu (1979: 85–87). On the development of technical prose, see Chapter 6.

of the south),[61] and the third on sericulture. Chen Fu says in his introduction that he wrote his treatise to refute the "empty words" and "irrelevances" of the two agronomic bibles of the time, the Northern Wei *Qimin yaoshu* and the Tang *Sishi zuanyao*, both written by northerners.[62] Unlike the authors of earlier compilations, Chen did not bother with quotations from earlier authors, pieties about the holiness of agriculture, or cosmological explanations of the processes he described. Every word counts in a text dense with practical detail, as in the following passage from Chen's chapter on growing rice seedlings for transplanting, entitled 'Taking care of the roots and shoots' (*shan qi genmiao pian*):

> When growing any crop the first thing is to take care of the roots and shoots so as to get the plant off to a good start, for it is very seldom that a crop finishes well if it starts badly. If you want healthy roots and sturdy shoots you must sow at the right time, meet the requirements of the land you have chosen, and use your fertilizers sensibly. If these three requirements are all met, and followed up by zealous attention, you will save yourself much worry and short of some natural disaster the crop will always do well.
>
> The first task when growing rice is to prepare the nursery-bed.[63] In autumn or winter it should be deeply ploughed two or three times so that it will be frozen by the snow or frost and the soil will be broken up fine. Cover it with rotted straw, dead leaves, cut weeds and dried-out stubble and then burn them off so that the soil will be warm and quick. Early in the spring plough again two or three times, harrowing and turning the soil. Spread manure on the bed.
>
> The best fertilizer is hemp waste, but it is difficult to use. It must be pounded fine and buried in a pit with burned manure. As when making yeast, wait for it to give off heat and sprout hairs, then spread it out and put the hot fertilizer from the centre to the sides and the cold from the sides to the centre, then heap it back in the pit. Repeat three or four times till it no longer gives off heat. It will then be ready for use. If it is not treated in this way it will burn and kill the plants. Nor should you use night-soil, which rots the young shoots and damages human hands and feet, producing sores

61 Water buffalo are not dealt with in the lengthy discussion of livestock in the *Qimin yaoshu*.

62 *CFNS*: 1. Both these works had quite recently been printed and circulated on imperial order, as mentioned in Note 47 above.

63 A small plot in which the rice is sown and tended until it is transplanted into the main field; this technique facilitates multiple-cropping since it reduces by six weeks or more the time that the rice crop spends in the main field. It also improves the quality of the crop since unhealthy seedlings are discarded, and the transplanting itself encourages root growth and tillering (the sprouting of multiple grain-bearing stems). A nursery bed typically supplies sufficient seedlings for an area of main field 15 or more times larger.

> that are difficult to heal.[64] Best of all the fertilizers is a mixture of burned compost, singed pigs' bristles and coarse bran rotted in a pit.
>
> The field should be soaked and brought to a fine tilth, then sprinkled with chaff and compost. Trample them into the soil, rake the surface quite smooth, and then you can broadcast the seed . . . I have often seen farmers use urine poured directly into the irrigation water; the resulting damage is immediately visible.[65]

Chen Fu's treatise describes a highly productive farming system in which rice cultivation and sericulture were closely integrated. Chen wrote as an experienced landlord about how to run a farm as an economic enterprise, providing detailed analyses of soils, criteria for matching available human and animal labour to land, and calculations of the inputs and profits of farming and sericulture. In the north, where millet crops might yield less than a tonne per hectare and where the short growing season precluded growing more than one crop of grain a year, 100 *mu* (roughly 6 hectares) was considered a reasonable size for a peasant holding. But in Jiangnan, double-cropping was already common in Song times, and it is interesting that in the passage on irrigation tanks quoted earlier Chen Fu gives an example based on 10 *mu* of paddy land, the same area that the seventeenth-century agronomist Zhang Lüxiang considered the maximum that a working landlord could manage.[66]

Chen Fu did not hold an official position like Lou Shou, and certainly did not have access to the ear of the emperor. His short treatise was not, like the *Gengzhi tu*, printed and circulated on imperial command. It did became extremely well known and influential, however, for it was incorporated almost in its entirety into the corresponding chapters of Wang Zhen's *Nongshu* (Agricultural treatise) of 1313, another great milestone in the history of *quannong* and of agronomic innovation.[67] Together these three works, the *Gengzhi tu*, and the *Nongshu* of Chen Fu and of Wang Zhen, depicted different aspects of the most advanced rice farming and sericultural techniques in Jiangnan at the time.

Chen Fu and Wang Zhen wrote in secular terms about how to foster the productivity of a given landscape while Lou Shou's scrolls of tilling and weaving were almost a rhapsody on the rural roots of empire. Yet the two formats, however distinct they may seem to us in retrospect, were complementary and inseparable

64 It is interesting to note Chen Fu's reservations about using night soil (human waste) as manure. For a history of the use of night soil in late imperial Jiangnan and its importance as an urban contribution to rural productive systems, see Xue (2005).

65 *CFNS*: 5–6.

66 *BNS*: 151.

67 See Chapter 6. Wang Zhen's *Nongshu* covered both northern and southern farming systems in great technical detail.

components of the imperial agenda of *quannong*. The landscapes that they depicted were both nourishing and instructive. Magistrates familiarized themselves with the latest best practice in order to teach the local people how to improve their farming methods and thus their livelihoods. They also organized lectures and rituals to remind the farmers of their dependence upon the state and their obligations to support the empire. If Chen Fu and Wang Zhen's landscapes represent a tradition of technical models, then Lou Shou's represent one of moral ideals, both of which contributed equally to the late imperial project of *zheng nong*, "correct agriculture".[68] The technical and moral dimensions of agricultural improvement were inseparable in the minds of Chinese officials – though perhaps not in the minds of farmers or landlords for whom the term *li* (as in *limin*, see above) was as likely to mean "economic profit" as "benefit" in the more general sense.[69]

A key dimension of the ideal landscape was its gendered nature. Many, though by no means all, early agricultural works included sections on the cultivation and processing of textile crops, and on raising silkworms.[70] Every household had to contribute yearly taxes paid in cloth as well as grain, in equivalent amounts, and state land allocations always factored in acreage for "hemp fields", *matian*, or "mulberry fields", *sangtian*, in addition to allocations for cereal crops. So it is not surprising that agronomists usually included instructions for producing fibre crops as well as grain, and for raising silkworms as well as horses or oxen. But their discussions were mostly limited to the production of the raw materials for textiles. One significant change ushered in by works such as the *Gengzhi tu* and Wang Zhen's *Nongshu* was that spinning and weaving, and all the other processes involved in the production of textiles, were given equal attention. Women and women's work were now literally in the picture. Both the *Gengzhi tu* and Wang Zhen's *Nongshu* devote as much space to the equipment and techniques of textile production as to those of crop cultivation, showing us an explicit gendering of space where men work outside in the fields and women work inside at the loom or in the silkworm shed: the ideal of *nangeng nüzhi* (men plough, women weave) had never yet been rendered so vividly.[71]

Once again, this iconographic and conceptual shift was triggered in part by the loss of the north in 1126. When the Song government fled south the key silk-producing region of Shandong was lost, and because so much silk was needed to

68 Will (1994).

69 Bray and Métailié (2001).

70 Wang Yuhu (1979).

71 See Kuhn (1987, 1988) for a more detailed analysis of textile imagery during the Song and Yuan. There were, of course, certain tasks where the boundaries between male and female work were crossed: men set up the bamboo frames in the silkworm sheds, old men helped the household women pick through the cocoons; and sometimes women were shown at work in the rice-fields (see Hammers, 2011).

establish and defend the new state, and to buy off the enemies threatening the northern borders, great efforts went into building up silk production in the south.[72] It turned out that the region around the new capital Hangzhou and along the lower Yangzi was ideally suited for silkworms as well as rice. Mulberry trees were planted densely along the edges of the rice fields and on sloping land, and this Jiangnan landscape, so richly productive of grain and cloth, sustained rapid economic growth while maintaining the gendered division of labour and the corresponding regime of taxation that were the foundation stones of the Chinese political and moral order.

How did this idealized meshing of landscapes of womanly and manly productivity translate into other environments? Some rice landscapes were integrated with sericulture: mulberries were planted along the bunds, their leaves fed silkworms and the silkworms' excrement was used as fertilizer. In other sericultural areas most of the land on the farm would be planted with mulberries and an area of less than one-fifteenth of a hectare would be planted with rice to support the family.[73] Cotton had become the most common textile crop in much of China by around 1500.[74] Unlike mulberries, cotton was a field crop that competed with staple grains for land, and in areas where farmers began to abandon rice to gain the maximum profits from cotton, local magistrates would often attempt to dissuade or prevent them. On the other hand, in areas where fibre crops were not well developed, officials encouraged their introduction. Figure 2.4, which shows women and children picking the cotton bolls, comes from a late eighteenth-century illustrated treatise on cotton cultivation, the *Mianhua tu* (Pictures of cotton), written by Fang Guancheng, the Governor of Zhili (the province that contained the capital, Beijing), to encourage the production of both raw cotton and homespun cloth. It contains detailed technical instructions suitable for the northern climate.[75]

Over the centuries following the Song, ideal landscapes of this type were reproduced in a steady wave rippling out from the heartlands of Jiangnan across the Chinese empire. It was official policy not only to spread the ideal rice landscape wherever in the empire it could be built (see below), but also to instruct local women in Chinese techniques for the production of yarn or cloth (see Chapter 3). Meanwhile in Jiangnan itself the methods and technologies of rice and silk and cotton production were intensified and improved,[76] and this new knowledge also tended to radiate outward, supplemented by information about technical

72 Kuhn (1987: 351 ff).

73 *BNS*: 101

74 See Chapter 6 on the introduction and dissemination of cotton cultivation, and its place in *quannong*.

75 Fang presented the manuscript version of the *Mianhua tu* to the emperor in 1765; a printed version was published under the title *Shouyi guangxun* (Expanded instructions on producing clothing) in 1808. For more on this work, see Kuhn (1988: 196–198) and Bray (2007b: 548–549).

76 Li Bozhong (1998a, 1998b); Shiba (1998).

FIGURE 2.4 Picking cotton, *Shouyi guangxun*, 1808. Cotton thinning and picking were among the few outdoor agricultural tasks (along with picking tea and mulberry leaves) considered appropriate for women, regularly performed despite their bound feet. This picture shows a group of younger women hard at work while a granny and a little boy, helpfully carrying a wicker basket round his neck, look on.

innovations or improvements from other regions. Official instruction and support played an important role in this diffusion of ideal landscapes; so too did migrating peasants, improving landlords and developing markets. I stress the term "ideal" because it is important to remember that these landscapes were considered symbolically as well as materially effective. Several scholars have noted that official attempts to introduce Han Chinese methods of farming and of textile production were part of a "civilizing mission", regarded as a highly effective means of sinifying barbarians and inculcating proper moral and political values. Owen Lattimore specifically points to the symbolic dimensions of official policies to establish Chinese farming landscapes along the frontiers.[77]

77 Lattimore (1962); Mann (1992a); Will (1994); Elvin (2004). Hostetler (2001) discusses the symbolic efficacity attributed to drawing maps of non-Chinese territories.

The rice landscape is not only highly productive, but also extremely versatile, and here technical and economic advances could generate tensions between official and farmer goals. Faced with the loss of the North, Southern Song officials taught and encouraged Jiangnan rice farmers to grow a second crop, of wheat, in the winter months. Many northern refugees preferred wheat to rice so there was a good market for the crop, yet initially farmers were reluctant, fearing that the practice would exhaust the soil's fertility and render their fields barren. But officials were able to provide information about recently perfected methods of fertilizing the fields, along with techniques for alternating between wet and dry cropping that improved the quality of the soil structure and cut down on pests. The government also wisely decided not to tax the wheat crop. In consequence, the Jiangnan rice farmers were soon converted to double-cropping.[78]

Many combinations of subsistence and commercial farming were possible in the rice regions. Depending on terrain, climatic conditions and prevailing markets, from the same rice field a farmer could sell the surplus of a single or double crop of rice, or grow catch-crops of vegetables or ginger or indigo. He could devote only the best land to rice for food, and use the rest for cotton or mulberries, oranges or sugar cane.[79] These levels of involvement with the market were seldom criticized by officials, nor was the sale of surplus cloth or yarn. But by the later sixteenth century, trade networks were so well developed that farmers in some areas were tempted to abandon any involvement in subsistence farming, and to concentrate all their resources on products for which they had a competitive advantage. Farming households around the prime silk region of Suzhou in Jiangnan, for example, preferred to turn over all their land to mulberries and to buy rice on the market. This is vividly illustrated in the *Bu nongshu* (Supplemented treatise on agriculture), completed in 1647 by the scholar-farmer Zhang Lüxiang who worked a smallholding in Tongxiang district just outside Huzhou, a region famous for producing the finest silk yarn in all China.[80] In the Shanghai area, on the other hand, many peasant farmers turned their farmland over to cotton.[81] The rice upon which they depended was produced upstream, in the middle Yangzi provinces of Hunan and Hubei, which had now advanced technically to the point where they were producing rice surpluses.[82]

78 Golas (1980).

79 Bray (1986: 124–134).

80 See Chapter 7 for more details on the *Bu nongshu*.

81 See Chapters 3 and 6.

82 Rawski (1972) is the classic study of this ripple dynamic, whereby regions that initially produced surpluses of cereals shifted to a more diversified economy of cropping and of manufacturing, while agriculture in hinterland regions that had been little involved in markets was developed to generate rice surpluses for export to the more advanced economic regions. See also Marks (1998a) and Pomeranz (2002).

In late imperial China, the lifeblood sustaining not only the state but also an increasingly intensive rural engagement in market production was the surplus of grain, and principally the surplus of rice. Some of this surplus circulated through local markets, some was transported across the country, whether in state barges moving tax grain north to the capital along the Grand Canal, or in merchant ships linking the fields of the latest grain-basket hinterland with the great cities of the Lower Yangzi. Twelve thousand official barges moved tax grain along the Grand Canal during the Ming dynasty. The standard load increased from 25 to 45 tonnes in the course of the dynasty.[83] Commercial shipping moved huge quantities of other commodities. For example, the Northeastern provinces were excellent for growing cotton, but the climate was too dry for successful spinning. Cotton dealers therefore transported raw cotton from the Northeast down the canal to the Yangzi delta. There it was processed and made into cloth, which was then taken north and sold by the same dealers.[84]

Although the inevitability of peasant involvement in trade was recognized in the great Ming tax reforms of the 1580s known as the "Single Whip", where all state labour obligations and taxes in kind were changed permanently to a single tax in cash,[85] the growing dependence of farmers on trade made late-Ming officials very nervous. They feared the risk of mass starvation in the Yangzi delta should the harvest fail in Hunan; they also faced ever-increasing needs for rice for official use, as enemies massed on the frontiers. In 1607, the statesman and agronomist Xu Guangqi compared the Chinese empire to a father with one industrious son (the rice-growing south) and one idle son (the north); the father, complained Xu, simply relied on the industrious son to support them all and thus encouraged the wastrel son in his idleness.[86] The late Ming government attempted to reduce the dependency of the capital and of the troops stationed along the northern frontiers by developing state farms; Xu Guangqi was prominent among the officials who experimented, fruitlessly, with introducing irrigated rice to the region. Such experiments were taken up again, to no avail, several times in the course of the Qing: 'The bureaucractic fantasy of turning dry northern fields into beautiful paddyland kept recurring despite the harsh realities of northern China.'[87]

It proved impossible to transplant wet-rice landscapes into the northern environment, and state officials would dearly have loved to insist that regions such

83 Brook (1998: 45–49).

84 Bray (1997: 212–225); Chapter 6.

85 See Huang (1974) for the Single Whip reforms and their background and impact.

86 Übelhör (1969: 44–47).

87 Lillian Li (2007: 89). Brook (2005c) provides a detailed account of these attempts to transfer and implant alien but essential agricultural and hydraulic skills and technologies. See also Chapter 7 on Xu Guangqi's experimentalism.

as Jiangnan devote themselves to producing rice to meet national needs. But there was no turning back the tide of commerce. By the seventeenth century the Jiangnan region had become a manufacturing powerhouse, propelled by household production. In Fujian, slightly to the south, some regions specialized in making paper or growing tea, while many Fujian and Guangdong peasants grew sugar, which was processed in small local factories using some specialized technology such as roller-mills.[88] Essentially, however, the main equipment needed for peasant sugar production was part of the same "technological kit" as that used for rice farming, and it could also be adapted without significant capital investment for other rural manufactures such as oil or soy products.[89] Smallholders found it relatively easy to switch between activities, and the rice-farming landscape served as the matrix for a thriving and versatile system of petty-commodity production. In China, commodities that were plantation crops in European colonies, such as cotton, tea or sugar, were grown on small peasant farms, processed domestically or in small local factories, and distributed by merchant networks throughout China and abroad. This system was often remarkably effective in assuring foreign markets for Chinese smallholders. For instance, well into the eighteenth century, when the rise of the Caribbean factories effectively displaced all competition, certain types of Chinese sugar were produced exclusively for European markets.[90]

Hamilton and Chang argue that this type of low-capital and highly flexible household involvement in commodity production allowed for extremely effective specialization and rationalization of production, with a mass consumer market emerging as early as the sixteenth century. The development through the late Ming of highly effective long-distance merchant networks was crucial in developing what Hamilton and Chang identify as a "distribution-driven" rather than "production-driven" market.[91] While acknowledging that in this system of trade-integrated handicraft production expansion often involved the simplification of technology and the increased involvement of younger family members, scholars such as Hamilton and Chang, and Pomeranz, reject interpretations of this trend as involutionary. Whereas Elvin and Huang have stressed the resistance to technical and managerial transformation and the potential for immiseration of a system of production that offered diminishing returns to an hour of labour, with peasants struggling to meet fixed consumption targets from shrinking farms, Pomeranz and other members of the California School put a positive spin on the flexibility of

88 See Gardella (1994) on tea, Chia (2002) and Eyferth (2009) on bamboo and paper production, and Sabban (1994), Daniels (1996) and Mazumdar (1998) on sugar.

89 Daniels (1996).

90 Sabban (1994).

91 Hamilton and Chang (2003).

this mechanically low-tech and labour-intensive system, arguing that late imperial China is a case of Hayami and de Vries' "industrious revolution".[92]

In 1968, Perkins had already argued that despite the population increases of the late seventeenth and eighteenth centuries, per capita supplies of basic foods did not start to drop in China until 1800 at the earliest. Pomeranz goes further, arguing that until the end of the eighteenth century much of rural China had high standards of living (measured by consumption of cloth, tobacco, sugar, tea and furniture as well as staple grains and fuel) that compared favourably to the most advanced areas of Europe at the time. He argues that one reason was that women's earnings more closely approximated men's than in Europe, and also notes that in China's rice-growing regions 'known techniques could still raise yields without exhausting the soil', while the preservation of natural resources such as forests also compared well to countries such as France, at least until the end of the eighteenth century.[93]

The ripple effect of economic intensification, population growth and raised consumption levels continued throughout late imperial times, filling up China's peripheries. The diversified economies of the core regions could support high densities of population and at first attracted migrants, but eventually a limit was reached and people migrated to less developed areas, which in their turn would see production intensify and population levels rise.[94] The Pearl River delta began to rival the Yangzi delta as a manufacturing core, importing raw materials and rice, and exporting processed goods.[95] The Shanghai region remained the chief exporter of cotton goods, and to meet its needs for raw materials it expanded its cotton fields while continuing to import raw cotton from the North. By the mid-nineteenth century the Canton region was importing raw cotton from India to feed its looms; it also sold sugar to the north in exchange for raw cotton. Starting in the eighteenth century, large quantities of rice were imported from Siam by

92 Hayami (1967, 1992); de Vries (1994). To explain what he sees as the flattening out of economic and technical development in China post-1400, Elvin (1973) uses the terms "involution" or "stagnation", while Huang (1990) speaks of "growth without development". The California School historians (see Introduction) prefer to use such terms as "Smithian growth" to characterize what they see as a historical trajectory of long-term if not revolutionary development. Historians such as Pomeranz place far less importance on *skill* as a substitute for capital or mechanical complexity than do I or Li Bozhong, or Francks in the case of Japan (Bray, 1986; Li Bozhong, 1998a, 1998b; Francks, 2014, in press).

93 Perkins (1968); Pomeranz (2002: 434).

94 The population of core regions such as Jiangnan remained stable through the Qing, while that of newly developing regions grew rapidly. 'The Yangzi Delta alone probably held 16 to 21 percent of China's population in 1750, but less than 9 percent by 1850 and under 7 percent by 1950' (Pomeranz, 2002: 440).

95 Marks (1998a); Cheung (2008).

various manufacturing regions of China.[96] Chinese merchant networks provided the empire with an extra-territorial "shadow acreage," which allowed it to compete successfully even in the late nineteenth and early twentieth centuries with many industrial products that Western nations had hoped to import into China, notably cotton goods.[97]

Glass half full, glass half empty

Let me finally return to the effects of this agronomic imperialism on natural resources. This chapter begins by documenting how and why the southern rice lands came to be perceived as a natural resource, displacing the northern millet-based farms as the ideal of a productive and socially harmonious landscape. It goes on to trace how the technical characteristics of intensive rice farming shaped a historical trajectory of economic, demographic and political expansion. The system was evidently very flexible, and in political and social terms it demonstrated long-term sustainability. But was it environmentally sustainable?

Clearly the long-term growth of population, coupled with rising levels of consumption and exports, translated into a steady increase of pressure on natural resources. To what extent were the rice and textile landscapes, and the socio-technical systems associated with them, able to accommodate these processes of economic growth without environmental degradation? Here again there are strong disagreements over the interpretation of the same data. Elvin, for example, presents levels of deforestation in 1800 as proof that from early times population growth and path-driven technologies had trapped China into a situation where it was probably more environmentally degraded than northwest Europe at the same time. Pomeranz interprets the same data as a long-term triumph of the Chinese political economy and its sustainable resource use.[98] Again, my own view of the historical potential of rice economies in general, and of the rice-based late imperial Chinese economy in particular, is closer to that of Pomeranz.[99]

What were the environmental effects of transplanting rice landscapes in order to extend the farming area and intensify cropping? This depended on the terrain and on the crop. The construction of new rice paddies had very different effects from clearing patches of hillside for maize.

Wet rice offers a comparatively stable basis for expansion and intensification. The shift to irrigated fields, transplanting, fertilizing and other intensive techniques not only increases output, but builds a highly resilient ecosystem underpinned by

96 Viraphol (1997).
97 Hamilton and Chang (2003: 199).
98 Elvin (2004); Pomeranz (2002).
99 See also Menzies (1994) on cases of forest protection and maintenance.

the enduring natural fertility and low erosion of established irrigated rice fields.[100] The long-term potential for intensification and diversification of crop production is very high. In 1178, the Song official Zhou Qufei wrote disapprovingly that the rice farmers of Qinzhou (southern Guangdong) were very careless in their methods, simply breaking up the soil into clods rather than working it till it was smooth, dibbling the seed into the field directly rather than transplanting, and neglecting to weed or irrigate – 'they simply leave Nature to take care of the crop'. By the late seventeenth century, Qu Dajun described how the farmers of the same region coaxed a continual sequence of rice, indigo, melons, oilseed or cotton from the soil, providing not only for their own subsistence, but also for large volumes of food and raw materials exported down-river to Canton: 'The people are all extremely industrious and devote themselves so diligently to their farming that truly no patch of land is wasted and no hands are ever idle'.[101]

Once the fields and watercourses had been built, wet-rice landscapes tended to be ecologically quite stable.[102] When rice farmers in the middle-Yangzi provinces of Hunan and Hubei began to export grain down-river to the Jiangnan cities, the intensification of rice production transformed the local landscapes but did not destabilize them. Rice yields could be raised by improved skills, increases in labour and cheap improvements in technique without exhaustion of the soil. Even poor farmers could afford the small purchases of soybean waste or oilcake that improved yields significantly if properly applied, and fertilizers were a trade that grew significantly through late imperial times.[103] Cotton, on the other hand, was known to be a gross feeder by comparison with rice, and officials continually nagged cotton growers not to neglect fertilizing the soil. By the mid-Qing the lower Yangzi provinces were importing huge quantities of soybeans from Manchuria not to eat but to keep their cotton fields in good heart.[104] In general the rice lands and surrounding croplands in the economically advanced regions did not deteriorate under pressure and remained highly productive throughout the imperial era.[105]

Not all Chinese crop systems were equally resilient, however, nor were all non-arable lands in the rice-growing regions necessarily well cared for. By the late eighteenth century, population increases and the consequent migration of many peasant families to more marginal and fragile lands in the central provinces and along the borders was starting to have devastating effects. Landless peasants moved up into mountainous regions, chopping down forests to plant crops that would

100 Grist (1975: 20 ff).

101 Bray (1984: 510, 509).

102 Osborne (1998: 208).

103 Bray (1984: 289–298).

104 Pomeranz (2002: 438).

105 See Shiba (1998). In the twentieth century the introduction of chemical fertilizers brought a quantum leap in productivity.

provide immediate returns on newly cleared hill-fields. The New World imports of maize and sweet potato were especially popular choices for poor peasants, while landlords invested in upland farms of tea or indigo. In both cases, the effects of deforestation and erosion, and the implications for China's river systems, were immediately clear to Chinese officials. Though they had no hesitation in legislating to curb landlord enterprises, control over remote areas was difficult to maintain. And in the case of the landless, compassionate magistrates were torn between allowing poor peasants to gain a livelihood, and protecting the waterways and fertile plains of China's most productive regions.[106] Until the mid-nineteenth century, officials in many of China's border regions also maintained 'remarkably effective government limitation of access to non-Han areas', thus protecting many fragile environments of grassland and mountains. Elsewhere, however, by the nineteenth century 'it seemed as if humans had set out to reduce their original natural environment to only two types of land: carefully maintained, productive, private farmland and ruthlessly exploited, unproductive, common wasteland'.[107]

Until the nineteenth-century crisis, official support for agriculture, in combination with the specific organization of Chinese trade networks, generally reduced the risks to farming families as well as local pressures on resources, while supporting increased production and consumption. The sheer scale of the Chinese polity, and its environmental and climatic diversity, also contributed. As Wong puts it, 'moving resources through greater spaces gave the Chinese state certain opportunities that European states lacked'.[108] Wong remarks that this often translated into lower taxation than might otherwise have been the case. It also allowed the government to dampen the effects of shortage and surplus in basic grains across its territories. In the course of the eighteenth century the effect of climatic disasters on grain prices was only a fraction in South China of what it was in England, and grain prices rose much more slowly, thanks to a constellation of intensive cropping patterns, an integrated national grain market, and state institutions to prevent hardship. The details of the latter were devised on the basis of a constant monitoring of food supplies where 'local officials routinely reported grain harvest estimates, rainfall amounts and grain prices to the central government'.[109]

Relations between state and merchants evolved through late imperial times. Wong notes that by the eighteenth century, although there were some official disagreements (animated by subsistence anxieties) over long-distance versus local commerce, or grain versus cash crops, generally the beneficial and important role of commerce was acknowledged. This, he notes, is a change from the sixteenth century, when the dramatically rapid expansion of commerce and perceptions of

106 See Osborne (1998: 222 ff) and Vermeer (1998: 273–275) for contrary examples.

107 Vermeer (1998: 278, 235).

108 Wong (1997: 132).

109 Marks (1998b: 422); Will and Wong (1991).

the loss of values and stability in pursuit of profit still generated widespread anxiety among the elite. But 'Chinese officials and thinkers came up with ways to tame commerce and create for it socially acceptable functions'. The state retained control over the bulk of the grain produced in China, though Chinese merchant networks imported additional supplies from Southeast Asia, and almost all other goods in daily use (apart from salt, which remained a state monopoly) were traded freely or subject to only nominal tax.[110] As mentioned earlier, the trade networks allowed for local specialization and regional complementarities (such as northern China and Jiangnan exchanging raw and woven cotton), thus also reducing the pressures on local resources while providing cash to farmers, some of which went into fertilizers, improved varieties or other inputs that improved productivity and kept the land in heart.

At the same time, as already mentioned, the ripple effect did not always work to support rational regional divisions of labour. In the eighteenth century, when the Middle Yangzi provinces turned to cotton instead of putting all their energies into rice, the surplus available for Jiangnan farmers dwindled and they were obliged to increase their own rice production – which in fact they did rather successfully.[111] Nor did the state's interventions to ensure stability, however well intentioned, always succeed. Elvin provides illuminating examples of several long-term large-scale hydraulic projects, which demonstrate that it is not only in modern societies that ambitious technological solutions generate new technical problems almost ad infinitum.[112]

Elvin concludes that the huge hydraulic systems he discusses locked economic and institutional structures into place, preventing any qualitative change, and he argues that large-scale state policies and local initiatives interacted in late imperial China to generate pressures on environmental, social and economic resources that grew inexorably until irreparable damage was done – a view in which Smil concurs.[113] Historians such as Pomeranz, Wong and Hamilton tend to emphasize the dynamic nature of the interplay between government, commercial networks and local economies, and they argue that state infrastructural support played a key role in improving factor productivity. Elvin alludes to 'a thoroughness of environmental exploitation that was distinctive in the premodern world'.[114] The revisionists (including me) would agree entirely with Elvin's formulation, while putting quite a different spin on it. The data are plentiful, if patchy, and new insights are offered every year, but China's very size, longevity and variety militate against a definitive demonstration that the glass was half empty or half full any time soon.

110 Wong (1997: 136–137).

111 Li (1998b).

112 Elvin (2004).

113 Elvin (2004: 162); Smil (1993).

114 Elvin (2004: xxiii–xxiv).

SECTION II

Gynotechnics

Crafting womanly virtues

In this section I retrace some of my steps in my earlier study *Technology and Gender*, looking at technical practices and meanings in three closely interlinked material domains (work, space and reproduction) that served in late imperial terms to define virtuous womanhood. The essays here engage more closely, however, with the different ways in which people might have mobilized the moral or emotional ambiguities of material convention. They also raise a number of questions about the significance in imperial Chinese contexts of material resources or habits that in European contexts have been identified as vectors of a modern economy or modern subjectivity, among them the commercialization of textile production, spatial privacy, and forms of fertility control.

3

WOMEN'S WORK AND WOMEN'S PLACE

Textiles and gender

I first came to study this topic because I was intrigued by a paradox. I had been working on agricultural treatises, dating from ancient to early-modern times, all of which quoted in their prefaces the classic phrase 'men plough, women weave', *nangeng nüzhi*. While the first half of a treatise of this kind deals with tillage, irrigation techniques and the cultivation of various crops, the second half covers the various processes of textile production. The illustrations to the first half depict men at the plough, transplanting rice, or threshing grain. The illustrations to the second half depict women tending silkworms, spinning cotton or weaving – either at simple looms or at complicated draw-looms, of the kind used for weaving patterned silks. I had vaguely taken the importance of women's work for granted. Then I read in the work of a modern historian that when foreign textile factories set up in the Chinese port cities in the 1920s, they employed women and the textile industry "became feminized".[1] This implied that by then textile production in China was considered "masculine" – so I wondered what had happened to change women's work into men's? And what were the consequences for women of losing control of a domain of production, skills and knowledge? How might this have affected gender roles and the status of women?

The Chinese themselves felt that material surroundings and physical experience played key parts in inculcating morality and forming social roles – an irresistible invitation to look for new ways in which ideology articulates with the material dimensions of everyday life. This chapter explores the values associated with a technical activity – weaving – which, because it involves the production of a key

1 So (1986).

commodity, has usually been analyzed principally from a modern Western economic perspective. This, I argue, has masked or veiled some of its most interesting dimensions, in particular the non-economic values attributed to spinning and weaving, and their significance as what I call "womanly work".

The saying 'men plough, women weave' expressed a doctrine of complementarity between men's work (farming) and women's work (textiles), predicated not only on a family's subsistence needs for food and clothing, but also on a moral contract between state and subject. From the pre-dynastic kingdoms right through to the late imperial period, the equal importance of male and female work was inscribed in a fiscal system that required every commoner family to pay taxes both in grain and in cloth. Husband and wife contributed jointly and equally to the support of the state, thus enabling it to provide for their welfare.

Between 1000 and 1700, complex changes took place in the textile industry that by and large displaced women as weavers; women were deskilled, and their contribution to textile production was devalued, marginalized, or simply subsumed within male-headed household production. What I want to discuss here is how such changes in the material domain might have played into a reconfiguration of gender roles.

Let me start with a very brief sketch of the main changes that took place in the Chinese textile industry between the Song and the early Qing.[2] Well into the Song all but a tiny proportion of the textiles produced in China consisted of simple weaves made by rural women in peasant or manorial households.[3] Not only was the workforce female: rural textile production was a *female domain*, in which technical knowledge as well as the responsibility for production were controlled and managed by women (Figure 3.1). The whole process of cloth production, from extracting the fibre to finishing the cloth, was usually performed under the same roof. Furthermore, although most of the textiles produced in this female domain were simple weaves of hemp, ramie or silk, the silk cloths in particular were not merely subsistence products but had a high recognized value, as commodities and as tax contributions. Silk tabbies, the plain under-and-over weaves that could be produced on the simplest loom, were used to pay taxes and as also served as a unit of currency.

In the Song an increase in commerce and prosperity and the growth of urban populations increased the commercial market for quality cloths, especially for fancy silks made on complex drawlooms. These looms were extremely expensive pieces of equipment requiring highly specialized skills, both of which lay beyond the reach

2 I provide a much more detailed account of how the organization of labour in the silk and textile fibre industries changed over the centuries in Bray (1997: Chapters 4–6).

3 In the Song, complex weaves made on drawlooms still formed only 4 per cent of total production even taking into account the most urbanized and advances silk-weaving regions of Sichuan and the lower Yangzi; Kuhn (1987: 378).

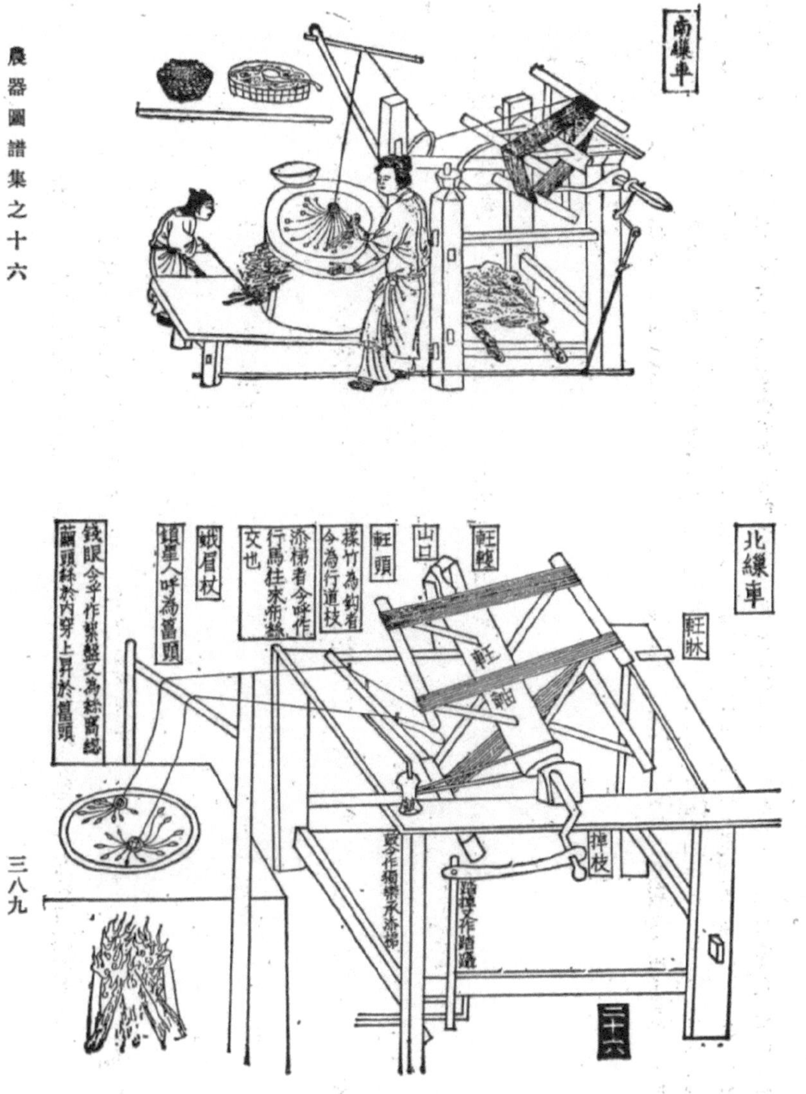

FIGURE 3.1 Southern (above) and northern (below) silk-reeling frames, 1530 edition of Wang Zhen's Nongshu. When Wang composed his work in 1313, northern silk technology was still more advanced than southern: the northern frame shown here reels two yarns simultaneously, the southern only one. The lower illustration names and explains the parts of the machine, but the upper illustration gives a clearer idea of how a reeling frame actually worked. The cocoons were floated in a pan of water heated on a stove. The hot water loosened the gum around each cocoon so the single long fibre could be wound off. The threads from several cocoons were fed together through an eyelet (the square hole in a copper coin), and as they emerged from the water they were twisted into a single yarn and wound onto a reel.

of ordinary peasant households. Urban workshops, which had previously constituted only a very small sector of the silk industry, grew in importance. These workshops depended on hired labour, and thus silk-weaving skills became increasingly a male preserve (Figure 3.2). The yarn the workshops used was purchased on the market and was mostly produced by women working at home, raising silkworms and reeling thread. These tasks were much less well rewarded than weaving, though male weavers too were exploited by their employers.[4]

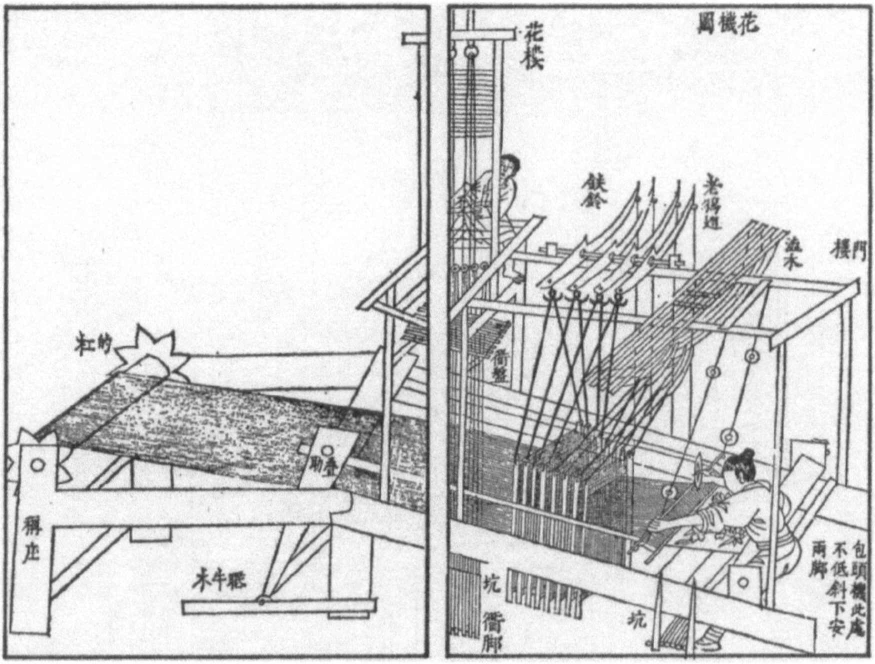

FIGURE 3.2 Drawloom operated by a man with a boy in the pattern-tower, 1637 edition of *Tiangong kaiwu*. This illustration was produced just a few decades before Jiao Bingzhen's more conventional reprise of the *Gengzhi tu* depiction of the drawloom (Figure 2.1), which still shows women operating the loom even though such practices were by now largely obsolete.

4 In Suzhou and the other lower Yangzi cities famous for their fine silks, a significant change occurred in the organization of weaving between the late Ming and the Qing. During the Ming the term "loom household" referred to employers owning workshops with several looms; they not only hired weavers but also paid workers to reel and twist the silk (lower-paid tasks that were usually performed by poor women, presumably at home), and to set up the looms. By the Qing, however, "loom household" referred to the weaver and his family, who hired the loom from the silk merchant and were paid piece-rates for working at home (Tong Shuye, 1981: 294). The merchant no longer had to hire reelers or other workers, since the weaver's family – his wives and daughters – now performed these tasks for no direct financial reward.

By the early Ming, silk manufacturers had to compete directly for markets with producers of fine cottons. The cotton industry had been marked by regional divisions of labour right from its inception (see Chapter 6), and this probably reinforced the trend towards regional and occupational specialization in the silk industry. During the Song, cotton was still a rare luxury; but under the Yuan, the Mongol rulers actively and successfully promoted the growing and processing of cotton, so that by the Ming most ordinary people wore cotton rather than hemp or other native plant fibres. Cotton grew better in the north of China, but the dry summers meant it was difficult to process there. A coarse cloth could be made suitable for peasant families' own use, but fine cloth, of the kind that competed with silks for softness, gloss, and design, had to be made with good-quality thread spun in humid conditions. Much northern raw cotton was sold to merchants who transported it to the market towns and surrounding villages of the Lower Yangzi, where it was spun and woven in peasant families. The merchants then dyed and calendered the cloth and re-exported it. In some rural cotton-weaving households along the Lower Yangzi the women wove the cloth as well as spinning the yarn; but it was quite common, in areas where holdings were small and farming profits slender, to find that the men of the household took charge of the weaving, while the women laboured to produce enough thread to dress the looms (Figure 3.3).

These economic changes produced political effects. Urbanization and occupational specialization gradually increased the number of taxable households that did not produce textiles themselves. During the Song, such households purchased textiles in order to pay their taxes; gradually, however, the state came to acknowledge the new economic realities and allowed for taxes in kind to be transmuted into cash payments. Families could choose whether or not to grow grain, or to produce textiles. Some regions such as Huzhou became nationally famous for their fine silk thread, while others with a long sericultural tradition gave up silk altogether. Some families grew cotton to buy rice, others bought the cotton cloth they needed with the proceeds of the sugar, tea or opium they produced. The fundamental break with subsistence production was recognized by the Ming state when it introduced the Single Whip tax system in 1581, finally severing the direct connection between the work of the commoner and the work of the state.

The overall effect of the changes I have described was to diminish the recognized value of women's work. Those women who did still spin and weave cloth at home did so usually for subsistence purposes, not for sale, nor for the payment of taxes. In the production of commercial cloth, where occupational specialization prevailed, women no longer held the honoured role of weavers but were reduced to the less respected and more poorly paid tasks of spinning and reeling. Even where whole families were involved in cloth production, again the man often sat at the loom while his wife and daughters served him, and he took the cloth to be sold. A class division also emerged. Formerly all women, whatever their social status, were

FIGURE 3.3 A simple loom, 1637 edition of *Tiangong kaiwu*. This is a waist-loom, that is to say, the tension in the warp is maintained by cords fasted to a belt around the weaver's waist. Waist-looms were popular for domestic weaving because they took up so little room. In the 1530 illustration of a waist-loom in Wang Zhen's *Nongshu* a woman is shown using this ubiquitous loom, but Song Yingxing has chosen to show a man weaving.

involved in making cloth. Elite women might not have sat at the loom themselves, but they organized concubines, servants and other junior women in the domestic production of textiles, a role that was recognized as contributing significantly to family wealth.[5] By late imperial times most gentlewomen were not involved in the production of cloth; instead they embroidered. These divergences from the canonical division of labour, where 'men ploughed, women wove', were just one manifestation of the complex social flux in late imperial China. Such changes were regarded by many with suspicion and unease, for they threatened a confusion of social roles, which would foster immorality and eventually result in chaos. By the

5 See the examples given by Kuhn (1988: 203) of Han noblewomen's organizational roles, and the admiring recognition given to their achievements in augmenting the family fortune.

mid-Ming, literati were already complaining about the dangers of the fact that men were weaving, so that women were no longer doing women's work.[6]

The masking or marginalization of women's economic contributions through new divisions of labour is of course not a uniquely Chinese phenomenon. It is a process familiar from many other societies in transition to a commercialized economy. In *The Origin of the Family, Private Property and the State*, Engels argued that in the process of social differentiation, which accompanied the development of the means of production and private rights to property, women became restricted to an unwaged sphere of reproductive activities distinct from a male sphere of socially recognized productive work. In this way women lost the status of full social adults to become subordinate wives and wards. From Engels on, Marxist and feminist scholars have argued that the reduced importance accorded to women's work under the pressures of commercialization and proto-industrialization has been a central factor in the consolidation of patriarchy around the world.[7]

But "patriarchy" takes many forms, even within a single society. To understand what these changes in the Chinese gender division of labour signified at the time, it is of course worth noting that they were paralleled in the proto-industrial textile industries of many regions of early, modern Europe, as they were in other parts of the world under the influence of colonialism.[8] But if we wish to use evidence from the Chinese textile industry to clarify the nature and history of Chinese patriarchy (or patriarchies), we must try to unravel the complicated tangle of meanings attached to "women's work" or "womanly work" in late imperial China – and this in the context of a historical trend towards commercialization and specialization in which the meanings of different kinds of men's work also changed, reflecting new social hierarchies and tensions. I shall consider two principal aspects of textile production in this regard: the specifically Chinese construction of the economic, social and political value of women's work; and the role of cloth in transcending the boundaries of gender segregation to link the inner and outer worlds of Chinese society.

I would like to signal at this point one noteworthy and initially surprising feature of the late imperial changes in textile production to which I shall return, namely the response they generated among powerful men. If the masking of women's productive contributions to the family economy served to strengthen the patriarchal power structure of the Chinese family, then one might expect that the Confucian theorists and legislators of Chinese patriarchy would accept such masking with approval, or at least let it go unmentioned rather than actively contesting it. But far from accepting the dwindling of women's recognized earning capacity as a proper mark of their subordinate position, from the emperor down, the educated male

6 Bray (1997: 242).

7 Sacks (1974).

8 Goody (1982); Weiner and Schneider (1989); Hafter (2007).

elite of the late Ming and early Qing were dismayed by what they perceived as an unnatural phenomenon that threatened the social order. In both their private and their public capacities they made numerous efforts (largely fruitless, I shall argue) to restore women's role as weavers. It is interesting to consider the nature of the policies devised and the reasons they gave for bringing women back to the loom, for restoring to women their work and the dignity which that work implied.

We are faced with an apparent paradox. The men at the apex of imperial China's patriarchal hierarchy did not want women's productive role to be masked within the family – they wanted it restored, made clearly visible, and honoured as a specifically female virtue. Susan Mann argues that such opinions relate to the growing and broadly recognized economic contribution that women in late imperial society made to the family finances.[9] In my view the elite discourses about "women's work" represented a more complex situation. Here I try to disentangle what the various writings or silences about "women's work" tell us about the tensions within late imperial Chinese society and about variant concepts of patriarchy and gender.

Womanly virtue and the preservation of the social order

When we try to understand what the loss of their previous identity as the possessors of weaving skills meant for the majority of women in late imperial China, we must think at once of the relations between family and state, of the roles attributed to family members, and of how distinctions were drawn between different social groups. We should also remember how closely body and morality were linked in Chinese thought. By performing wifely work a woman practised both family and civic virtues. When a little girl was taught to spin and weave, she was not only acquiring the skills to produce useful goods – through them she was learning diligence, orderliness, and respect for labour, the dignity of a wife and the responsibility of a subject of the state. In the cloth that she produced, her skills were transmuted into worth and virtue. That is one reason why so many social reformers and moralists of the late imperial period were anxious to revive women's weaving skills.

When women abandoned proper women's work, the whole fabric of society was threatened. As early as the fifteenth century the writer Hu Juren (1434–1484) was lamenting that commercialization and the spread of artisan activities meant that weaving had now become a man's job.[10] In Confucian terms this was a threat to the social order in part because basic production had fallen into the hands of artisans and merchants, but equally because it signalled a failure to distinguish properly between male and female work. Men weaving or women tilling the fields were deeply disquieting occurrences, because they were signs of a world turned upside down.

9 More recently, Ken Pomeranz (2003) has elaborated this argument.

10 McDermott (1990: 30).

Where the social order of the state forms a continuum with the conduct of family life, women's behaviour is a matter of political concern even though they live in seclusion from what we might think of as the public sphere. The 1712/13 preface to Lan Dingyuan's popular text *Women's Learning* voices a typical opinion when it declares:

> The basis of the government of the empire lies in the habits of the people. The correctness of the habits of the people depends on the orderly management of the family. The Way (*dao*) for the orderly management of the family begins with women.

As early as Han times the female scholar Ban Zhao had defined four attributes that all women should cultivate: womanly virtue, womanly speech, womanly conduct, and womanly work.[11] "Private" virtues were considered to underpin the "public" order in many ways; and for late imperial moralists, women's weaving incorporated several levels of virtuous behaviour.

Although in reality most elite households by the Ming and Qing had long since abandoned cloth production and supplied their needs through the market, popular encyclopedias as well as agricultural works of the period continue to depict spinning and weaving as the daily occupations of respectable ladies (for example, Figure 2.1). One significant way in which texts of this kind represented domestic textile work was not as a commercial activity, but as an exercise in *xiao* (filial piety or respect for elders and superiors): in several *nügong* series the final picture shows women presenting the warm and delicate cloths they have woven to the patriarch and matriarch, to choose their new clothes for the coming year.[12] Spinning and weaving taught both the gender-neutral virtue of filial piety, and the more specifically female virtues of thrift, frugality and diligence. These were considered essential not only for poor households but also for the proper management of elite households, in which parents still admonished their daughters to spin in order to learn respect for the hard work of their inferiors, and to weave hemp to learn frugality. The eighteenth-century conservative Chen Hongmou wrote bitterly of spoiled young women whose children were raised by nurses, while maids and concubines did their needlework.

> All they have to think about is making themselves beautiful. Everything is done for them, so they don't know that rice comes from a stalk and silk is unreeled from cocoons. They treat money like dirt, and living creatures like bits of straw.[13]

11 Lan Dingyuan's *Nü xue*, quoted in Mann (1994: 23). On the history of the four attributes and their use in late imperial didactic texts for women, see Mann (1994: 21 ff).

12 Bray (1997: 258).

13 Mann (1991: 220).

Most educated people in late imperial China saw idleness as a danger – and as wealth grew, more and more women were perceived as being idle and therefore under moral threat. The sign of being idle, in orthodox eyes, was that the women did not produce cloth. The late Ming and early Qing literature provides quite a few examples of official attempts to recreate a textile industry in poor regions where women no longer wove.[14] These women were described in the proposals for reform as being "without occupation" or some such expression – even though they were probably working hard all day long, and may well have been involved in farming tasks, handicrafts or other economic activities as well as housekeeping and child-rearing. The only women's work that was worthy of that name in the eyes of the reforming officials was making textiles. On the other hand there was a growing number of well-off households whose wives and daughters did now have what we would recognize as leisure time, time they might occupy with reading, painting, or embroidery (a skill to which I shall return later). Such leisure was considered especially threatening to women of the commoner class. As well as attempting during his posting there to revive the silk-weaving skills of the poor women of Luzhou, an impoverished city in Shaanxi province that had once been famous for its fine tabbies,[15] the sixteenth-century official Lü Kun composed a work entitled *Gui fan* (Regulations for the women's quarters) specially to provide moral instruction for commoner women who now had 'three or five books in their chests' but less work than was good for them.[16]

In an essay on 'Household handicrafts and state policy in Qing times', Susan Mann provides a series of examples of official efforts to eradicate peasant women's idleness and its attendant dangers by encouraging what she terms "handicrafts" – though it turns out that we are dealing here not with a whole range of acceptable handicraft activities, but invariably with textile production and especially home weaving. In regions that were not yet fully sinicized, teaching Chinese-style weaving skills was seen as a powerful method of enculturation. Mann quotes a touching lament by an official posted to the Xiang region of Hubei, where the local women (presumably of some other ethnicity than Han) were shamelessly unfaithful to their husbands by Chinese lights. They thought nothing of marrying several times, nor did families in search of a bride mind this loose behaviour.[17] The trigger to the

14 See Will (1994), Rowe (2001) and McMahon (2009) on successive Qing attempts over more than two centuries to revive the silk-tabby industry of Shaanxi.

15 See Bray (1997: 228) on the late-Ming demise of the Luzhou silk industry and the impact on regional sericulture.

16 Handlin (1975: 17). Lü Kun was a prominent advocate of female literacy even for commoners, arguing that female education would strengthen the complementarity of the husband-wife relationship.

17 This was a period when there were strong moral pressures on Han women to remain faithful to their husbands by refusing to remarry. However, widows often came under severe pressure to marry again, not least because of legislation that allowed the in-laws to keep the dowry if the woman remarried (T'ien, 1988; Carlitz, 1994).

official's concern was the sight of the women 'breaking up clods of earth with a hoe'. This shocked him deeply since such a confusion of gender categories indicated that 'it would not be long before these same wives would be suing their husbands, and husbands their wives'. Only sericulture, he felt, could save them:

> I especially pity the women of Xiang. They have no way to develop specialized work of their own and thereby affirm their commitment as faithful wives. The woman who has no work of her own should take up sericulture. Whether she comes from a gentry or a commoner household, a wife can personally tend silkworms in order to clothe her husband. When she sees that her own strength is sufficient to provide for her family's subsistence, her heart will be pure.[18]

Within the traditional Chinese heartland too, Qing statescraft writers believed that 'those regions in China where "women wove" were culturally and even morally superior to areas where women [did not]'.[19] Chen Hongmou was one of many Qing officials who tried to introduce the skills of advanced areas such as the lower Yangzi to more backward areas where they held office, even though the local population would sometimes refuse to learn them.[20]

One reason why local populations did not always welcome these attempts was precisely because they had developed alternative and more rewarding means for earning a livelihood. In another article Mann documents the real range of home handicrafts that women of various classes practised in the Ningbo region during the late Qing period. They included embroidery in well-off families, and also 'handwoven cloth, rush mats and matting, hats made of straw or bamboo splints, and the canopies of oiled-paper umbrellas. All were counted among the specialities that made Ningbo famous' throughout Southeast Asia as well as China.[21] But these were not the type of women's work that late imperial officials were likely to encourage or promote, for they were activities that fell squarely under the category of craft, and were therefore – unlike textile production – considered not basic but subsidiary and commercial activities. Even in the Qing, officials still tended to treat commerce as a drain on the state's resources and the people's welfare. When late imperial officials encouraged home weaving or sericulture, they did so for two

18 Mann (1992a: 87), quoting from a collection of essays on agricultural policy that comprise three chapters in the *Huangchao jingshi wenbian* [Collected essays on statecraft of the August Dynasty], first published in 1826. 'More than half of the forty-nine essays in these chapters refer to home weaving, and one entire chapter containing nineteen essays (*juan 37*) focuses exclusively on cotton and silk home handicraft industries' (Mann, 1992a: 92, Note 8).

19 Mann (1992a: 86).

20 See Will (1994) and Rowe (2001) on Chen Hongmou, and the references in Chapters 2 and 8 to Fang Guancheng's efforts to develop cotton manufactures in the capital region of Zhili.

21 Mann (1992b: 249).

reasons: first, it allowed women to contribute to their family's subsistence needs, and second, it meant they could pay their taxes.

Late imperial officials and moralists did not neglect the economic value of women's work, but they reduced it to its political elements: family subsistence underpinning a moral contract between subject and state through the payment of taxes. The same was true of their attitudes towards men's work. It was not that they objected to modest prosperity, and many felt that a moderate involvement in commercial production or trade was legitimate and even desirable – so long as it was firmly underpinned by subsistence production. Qing emperors and officials did all they could to coax commoner men out of trade and crafts and over-reliance on commercial cropping, and back into farming based on essential grain production. In their minds such agricultural policies served as a fetish: the restoration of "correct agriculture", that is to say subsistence-oriented farming, would not only re-establish a firm tax base for the state but would also revive the ancient social contract between ruler and subject.[22] I see late imperial official efforts to restore a traditional textile industry as representing the female side of the gender equation underpinning this moral order. Traditionally both man and wife had been responsible for paying taxes – both had been active subjects of the state. Weaving, like grain farming, served as a metonym: it represented not just the private female virtue of "womanly work", nor an essential element of the family's subsistence, but also the necessary female contribution to a desirable public and political order.

It is ironic that the lower Yangzi was the region most often held up by officials as an example of proper women's work. Although it was true that in the sericultural regions around Huzhou even the poor (or most of them) could afford to pay their taxes, by the Qing the local economy was definitely not based on self-sufficiency but on a highly developed market economy (see below). The paradox is even clearer in the case of the cotton industry. Even in highly commercialized regions such as Fuzhou, a thriving centre for tea, paper and printing, officials persisted in pressing farming families to grow cotton and weave their own cloth. Yet most local women were profitably involved in processing tea, tobacco and sugar and could very well afford to buy cloth on the market. At the very same time officials in the lower Yangzi were trying to stop farmers specializing in growing cotton because they did not wish them to neglect grain, even though the farmers told them that cotton was far more profitable than rice.[23]

As far as Confucian officials were concerned, modest self-sufficiency was infinitely preferable to the kind of wealth that generates what we would call economic growth, for the latter involved the risks of dependence on the commercial economy. One doubts whether this view was often shared by commoner families

22 See Will (1994: 875–882) on "correct agriculture" and the preservation of the social order.
23 Mann (1992a: 84–85).

in the late imperial period. It is true that where transport networks were not well developed, abandoning subsistence production made a region vulnerable to famine. But that is not all that was at stake, I think, when officials urged the local farmers to stop growing so much cotton, or sugar, or oranges, and turn their fields over to rice. Nor was it even true that taxes on subsistence farming were more reliable or easier to collect than taxes on other livelihoods after the complex reforms of the late Ming. Something else lay behind the dogged economic conservatism of the Qing officials. As Hill Gates puts it, the late imperial state did all it could to maintain a 'tributary mode of production' in the teeth of the consolidation on the ground of a 'petty capitalist mode of production'.[24] It was as if the late imperial officials who held up the lower Yangzi cloth industry as an example were deliberately blind to what was really going on – cloth was being produced and taxes were being paid, and from this they deduced that the women of the region were fulfilling their traditional role, even though the articulations of the lower Yangzi cotton industry had been completely transformed.

However, the sericultural industry of the same region did develop in such a way as to fulfil both the moral expectations of officials, and the material needs of the local population. Sericulture, the hatching and raising of silkworms in order to produce cocoons, was one area of textile production that was an exception to the deskilling and marginalization of women, remaining a female and domestic domain of competence through all the specialization and sub-divisions of textile production of the late imperial period (Figure 3.4). In a sense, the splitting-up of the various silk-making processes (into sericulture, the reeling and quilling of yarn, and the actual weaving) actually enhanced the importance of raising silkworms – it was so clearly the rock on which the silk industry was founded, and silk thread had always been a valuable commodity in its own right.

By the late Ming, in certain regions of the Lower Yangzi such as Huzhou, where the best cocoons were produced, the whole economy depended on sericulture, and dry land on which mulberries could be grown was much more valuable than rice fields. 'In my village', writes Zhang Lüxiang, 'the profits from sericulture are greater than those from grain, and the public and private sector[25] alike depend on them. If the silkworms don't spin well then public and private alike suffer terrible hardship'.[26] Zhang insists upon the absolutely essential contribution that the wife makes to the domestic economy, both through her productive and her managerial skills:

> If a woman works diligently, her family will flourish, and if she is negligent, it will decline – exactly as with a man's tasks. Because a woman's occupations

24 Gates (1989); see also Will (1994).

25 The public sector means the official administration, but also presumably community institutions such as local schools, temples and charities.

26 *BNS*: 108.

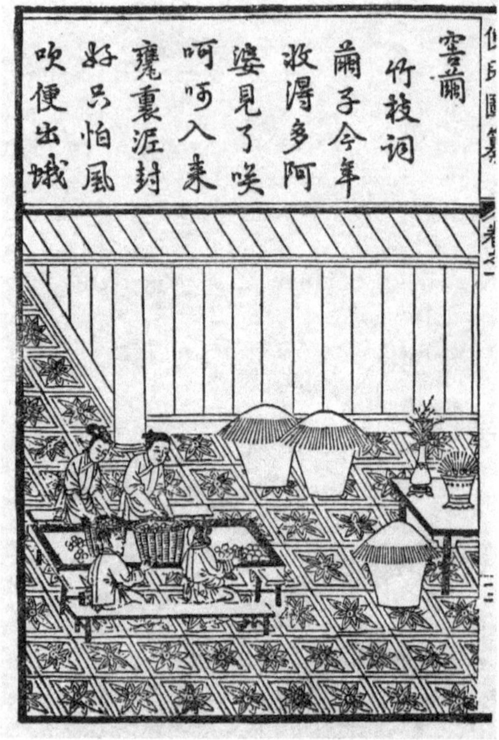

FIGURE 3.4 Storing cocoons, *Bianmin tuzuan*, 1593 edition. After the cocoons had been sorted, they were treated to kill the chrysalis and stored ready for reeling. This picture shows the straw-covered pots in the background, in which cocoons were packed with alternate layers of pauwlonia leaves and salt to kill the chrysalis; a more common method was steaming the cocoons.[27]

> are limited to such things as hemp or silk, it seems as if her diligence or laziness would have only a slight effect on the way the household runs. But in fact if she is diligent then all the tasks will prosper and if she is lazy they will all be neglected. So they say, 'A poor family hopes for a wise wife as a state in disorder hopes for a good minister.' The degree of help and cooperation is the same in both cases.[28]

Zhang presents the wife as an active and equal partner in the domestic economy:

> Of husband and wife, the man can manage 10 *mu* of land, wet or dry, and the woman can rear ten frames of silkworms. In one day she can weave

27 Kuhn (1988: 343).
28 *BNS*: 151.

> 2 bolts of ramie cloth, or spin 8 ounces of cotton yarn – how could such a family go cold or hungry?[29]

Zhang notes that a skilled and diligent wife can even reverse the family fortunes. More than that, he tells us that the local people compared a wife's role to the dignity and responsibility of a minister of state. This takes us even beyond the formulation in the *Book of Rites* that 'the wife is the fitting partner of her husband'.[30] In sericultural regions not only did the household depend on female sericultural skills but so did the whole local economy, including the public institutions funded by tax money.

Huzhou sericulture was not in fact a traditional subsistence activity. Only a small proportion of the silk thread was actually woven within the household that produced it (and in such cases the men often did the weaving while their wives and daughters span and reeled the thread). Instead it tied sericultural families into a complex commercial chain that linked them to weaving workshops in Suzhou or Hangzhou that produced cloth for the national market, or even into networks that extended far beyond the provincial borders through the national and international manufactures of Nanjing and Canton. However, as producers of cocoons and yarn, women were still at the centre of this domain of traditional "womanly work". State ideologues chose to emphasize that this work allowed sericultural families to subsist in dignity and to pay their taxes; when they offered it as an example to other regions they simply ignored the commercial networks that played on comparative advantage, allowing this form of sericulture to flourish in the Yangzi provinces where the silk thread was especially fine. The work was hard and the rewards slender even in Huzhou where the best quality silk thread was produced.[31] Little wonder that few households in remote areas such as Luzhou were tempted to persist with silk production once the eager official who devised the project had been transferred to a new post.

Despite the realities that restricted sericulture in Ming and Qing China to a few favoured regions, to orthodox thinkers it still represented a very powerful symbol of the traditional social order, and that, I think, is what lay behind the Qing revival of medieval court rituals in honour of the goddess of sericulture. This ceremony, in which the empress and her ladies tended silkworms in a special shrine at the capital, was first developed during the Han dynasty along with the parallel "first furrow" ritual in which the emperor himself set hand to the plough. Although it could be argued that these rituals were simply intended to ensure good harvests of grain and of cocoons throughout the empire, I think they were equally

29 *BNS*: 151.

30 'The wife is the fitting partner of her husband, performing all the work with silk and hemp' (Kuhn, 1988: 20).

31 See, for instance, the calculations of wages and prices in *BNS*: 84.

concerned with the reproduction of a moral and political order, serving as symbols of the interdependence between emperor and subject, and of the ruler's respect for the fundamental occupations that defined and supported the Chinese agrarian state.[32]

Although the court sericultural rituals performed by Qing empresses were careful reconstructions and elaborations of much earlier ceremonies, the popular industry they were intended to promote had changed almost beyond recognition. If conservative officials and elite moralists still clung to such comforting shreds of gender tradition as Huzhou sericulture that could be proposed as a model for regeneration elsewhere, it seems that for most ordinary people the understanding of woman's work and its importance must have changed considerably. Now I want to discuss how women's productive work related to ordinary understandings of man and woman, and in particular of husband and wife.

Women's work and family status

The marginalization of women's productive roles is a rather typical occurrence in societies in the process of proto-industrialization. But only by examining the broader context in which such changes take place can we understand how they might contribute to reconfigurations of gender roles in specific instances. In the case of medieval Europe, for instance, David Herlihy argues that the rise of male guilds and the diminished authority of women in the textile industry did not in themselves account for the parallel decline in women's social status, rather, changes in marriage and inheritance practices were the primary factors.[33]

In late imperial China likewise, changes in women's legal status and in kinship practices defined a matrix that structured the effects of changes in women's work roles. With the replacement of an aristocratic governing class by an elite of scholar-officials, the early Song heralded a shift in marriage exchanges from bridewealth to dowry. Neo-Confucian formulations of orthodoxy now excluded daughters from rights to a share of the patrimony and to inclusion in their natal lineage.[34] Instead wives became recognized members of their husband's lineage once they were mothers to a son. In this context it does seem that the erosion of women's productive capacities – or at least of their acknowledged productive capacities – provided fertile soil for a growing emphasis on the reproductive roles of wives: as childbearers,

32 Kuhn provides a historical survey and analysis of court and local sericultural rituals (Kuhn, 1988: 251 ff). A detailed description of the Qing ceremonials, which were particularly lavish, is given in (Mann, 1992a: 80–81). See also Bray (1997: 250–251).

33 Herlihy (1985: 100–102).

34 Ebrey (1991c) illustrates how criticism of buying brides (as a way of gaining entry into aristocratic circles) gave way in the Song to complaints about buying grooms (or at any rate those grooms who were reckoned to have good minds and a brilliant career in government ahead of them).

and as managers rather than generators of the household economy.[35] Moreover, one can see how these conditions taken together could translate into the view that women consumed without producing, an ethos that at least at popular levels could develop into negative perceptions of daughters and even of wives.

Dowries, for example, were less of a burden when women were still active in weaving; at that time they not only contributed to tax payments and earned extra cash, but were able themselves to manufacture large portions of their own dowries. By late Ming times official reports on the distressing phenomenon of female infanticide inform us that most peasant families had difficulty providing dowries, and the problem was compounded by the fact that daughters did not – could not – contribute to household earnings. The view that daughters were considered a burden was reflected in innumerable popular sayings about the worthlessness of girls.[36]

Hill Gates sees the changes in textile production as one factor in the 'commoditization of Chinese women', in which the patrilineal, patriarchal Chinese kinship system permitted men to appropriate the fruits of women's economic and reproductive labour.[37] Gates argues that the unprecedented economic growth of the Song provided an unprecedented range of new opportunities for women as producers and earners of income. She hypothesizes that this perceived threat to male control underlay the misogynistic severity of the neo-Confucian doctrines of gender hierarchy elaborated at the time, as well as legal changes that strengthened the control of the male-centred lineage over any female claims to property.[38] The family economy operated within the overarching rationalities of a market-driven 'petty capitalist mode of production' in which households or patrilines strove to increase their incomes, or – in the case of poor families – struggled just to survive. The interests of these private economic units, says Gates, frequently clashed with those of the state, which operated in the interests of a 'tributary mode of production', in which a ruling class of officials extracted taxes from the great body of commoners. Despite its secular growth throughout the late imperial period, its flexibility and its sophistication, the petty capitalist mode of production never came to dominate the Chinese economy, argues Gates, because the state always acted to contain it.

35 This should not be taken to mean that for all women biology was destiny: the legal wife in polygynous households was the social, ritual and legal mother of concubines' children, and adoption was by no means uncommon; see Chapter 5 and Bray (1997: 335–368) on 'Reproductive hierarchies'.

36 T'ien (1988).

37 Gates (1989, 1996).

38 A survey of references to women's occupations in Song sources yields rather sparse evidence to support Gates' view. There are only a few mentions of women running tea-shops, herbalist stores, or other small shops, and of course also to women earning at home-based textile production as well as tailoring, sewing and embroidery; otherwise, the chief occupations mentioned are the perennial ones of maid, kitchen maid, and various kinds of entertainer or prostitute (Jin, 1988).

Although I feel misgivings about Gates' interpretation of the relations between the economy and the birth of neo-Confucian gender doctrines during the Song period, her particular formulation of two economic systems in continual tension[39] helped me unravel some of the contradictions I saw in the attitudes towards women's productive work in late imperial times, contradictions that seem to me to crystallize in the ambivalence of the term *nügong* (or *nühong*).[40] Throughout this chapter I have alternated between translating *nügong* as "womanly work" and "women's work", and in the course of writing I came to realize that this is in fact a very illuminating distinction. Of course it is not a distinction that a Chinese speaker or writer needed consciously to make, hence some of the confusions past and present in dealing with what is apparently a single category.

"Womanly work" is the translation I decided to use (following such scholars as Susan Mann and Dorothy Ko) principally to denote moralists' and officials' use of *nügong*. What these writers see in female work is a moral activity linked to a gendered identity, and embodied in weaving. This notion of work corresponds to Gates' tributary mode of production. It implies that the woman is an active subject of the state, and that her role is an essential complement to that of her husband. The work performed is the production of symbolically charged cloth, an essential good like food grain; the straightforward monetary value of the commodity produced by this work is of no real interest. This is why late imperial officials were able to describe women who did not weave as having "no occupation", and why they were able to see in the highly commercialized sericulture of Huzhou an exemplar of traditional wifely occupations.

The translation as "women's work" I see as operating at the level of the private household economy, within the framework of a market economy of the type that Gates calls petty capitalist mode of production. In this context, *nügong* could mean any kind of work women performed that produced recognizable commodities, ranging from picking and processing tea[41] to the kinds of handicraft that a Ningbo survey of 1907 documented as women's home-based work:[42] weaving mats, making umbrellas or hats, processing foods, all kinds of work that fit very nicely into our Western concept of handicrafts, but which as far as Confucian orthodoxy was concerned did not qualify as "womanly work". Modern scholars such as Tong Yeshu, Philip Huang and Susan Mann tend to categorize weaving as a "handicraft",

39 The contradiction between state goals and private interests is, of course, a rather tired commonplace of Chinese economic history. By bringing gender *and* class into the equation, however, Gates opens up new possibilities for dynamic analysis.

40 The second character (pronounced in this combination either as "gong" or "hong") can be written in three different ways.

41 See Lu (2004) for an analysis of the interplay between "women's work" and "womanly work" in the tea-picking industry in South China during the Ming and Qing.

42 In 1907 a native of Ningbo, Nyok-Ching Tsur, conducted a survey of Ningbo industry (1983 [1907]), which Mann (1992b) uses as one of her key sources.

thus eliding a distinction between basic and subsidiary activities that was absolutely fundamental to orthodox Confucian economic thought, but that probably became alien to most ordinary working people as commercialization advanced and specialized divisions of labour became the norm, not only in the textile industry and in crafts and trades, but also in the canonical male activity of farming. As work of all kinds came increasingly to be evaluated in terms of cash wages or prices, and as working people scrambled from one crop or one occupation to another in order to survive, the idea that work was a dignifying sign of a particular social identity must have faded and eventually disappeared.

Reforming officials and moralists in late imperial China, I think, held to the hope that "women's work" could be transformed into "womanly work", with all its moral powers and its implications for the revival of a more desirable social order. But most ordinary people in late imperial times, struggling to survive in a commercialized economy by any occupation that offered a living, probably shared our modern view that weaving was an occupation like any other: it was of little importance whether the female members of the household wove cloth, cured tea, painted fans or brewed soy sauce, provided it brought in some cash. Although there is no systematic popular equivalent to the elite discourses on the value of *nügong*, I suspect that in working families in an increasingly market-driven economy, one factor that powerfully affected women's status vis à vis their male kin was their *perceived* ability to contribute to the family finances.

Philip Huang remarks that the commercialization of the Chinese peasant economy in the late imperial period actually increased the involvement of all family members in production, including women.[43] This may seem to contradict the arguments I have been making about marginalization – but it does not, for Huang is talking about the facts of economic activity, and I am interested in its representations.

In attributing responsibility for a particular kind of work, carrying out the final transforming process is often what counts. In China, whoever sat at the loom was considered the real maker of the cloth. Cocoons and silk thread were valuable commodities considered to be the product of women's work. Perhaps the same was true of curing tea or other processed or handicraft products where women were responsible for the key transformation. But more often the splitting-up of production processes that accompanied the commercialization of the late imperial Chinese economy not only produced new and specialized divisions of labour and increasing degrees of exploitation, it also deprived women of primary responsibility for various products, and rendered invisible the value of female work. For instance,

43 Huang (1990: 52, 54). Huang interprets these trends to support his argument that the late imperial economy was involutionary. Pomeranz (2003) revisits many of the same sources on women's work to argue instead for a long-term trend of Smithian growth.

in a world where most goods had an explicit cash value, if a woman wove cloth for family subsistence needs, what she produced was only attributed a proper value in cash terms if she failed to clothe her family and extra cloth had to be bought. The masking of the value of home-made cloth was reinforced by the tax reforms that made cloth payments obsolete. And women who worked in the commercial production of cloth were either the poorest paid workers in the new divisions of labour between households and workshops, or they worked as the auxiliaries in a household team where the men wove the final product. In this system both men and women workers were exploited and alienated, but men were usually allocated the more remunerative work. Women who produced cotton yarn for sale, or who worked as silk quillers or reelers, were paid mere pittances; the male weavers or weaving families worked under slightly better conditions, though as I argued earlier these tended to deteriorate with the development of the market economy. So in many families the value of women's work was either perceived as less than that of men's work, or it was rendered invisible, appropriated (as Gates says) by their male kin.

In the cases where women were still seen as primarily or solely responsible for the production of a valuable commodity, the value of the work they performed does seem to have translated into respect. Women who practised sericulture and those who wove cloth for sale were respected by their families and by the community, as well as being honoured for their womanly qualities by officials. Zhang Lüxiang's account of the positive image of peasant wives in late Ming sericultural families is one example of how people who visibly earned money for the family were well regarded whatever their sex, even at a time when neo-Confucian orthodoxy about female subordination had filtered down to the lower levels of society. In the sericultural families that Zhang describes, the role of a wife was compared to a minister, and the relationship between spouses was represented as one of complementarity and interdependence. It would be interesting to know whether women who cured tea or made handicrafts were similarly regarded.[44]

Having argued thus far that "women's work" rather than "womanly work" was the issue at stake in ordinary working families, I should also point out that even in working families "women's work" was delimited in certain ways that held the moral implications of "womanly work". Though within the walls of their home women might perform a wide range of tasks that generated income without compromising their reputation, women who worked outside the home lowered the status of their family. Female seclusion was a basic mark of respectability and

44 Lu (2004) documents elite male appreciation of the economic contributions of tea-picking women, but evidence of popular opinion is more elusive. T'ien (1988) tells us that where outside employment opportunities for girls were frequent – the examples he gives are singers and prostitutes – rates of female infanticide were much lower among poor families than elsewhere. But we know much less about more respectable forms of employment within the household.

orthodoxy that every family hoped to retain, and only the poorest and most desperate families would allow their women to go out to work.

This brings up the interesting question of inner and outer, private and public spheres. How should we understand such concepts in the context of late imperial China, and to what extent did female seclusion translate into a separation of worlds? I would like to argue that most work tied women into the outer world, but that "womanly work" implied different kinds of ties from "women's work".

Connection and seclusion: cloth and the separation of spheres

Engels' thesis about the relation between socially recognized work and social status underpins a range of Marxist and feminist analyses of gender hierarchies, which circle warily around the hierarchies implicit in the distinction between "domestic" and "public" spheres. Fruitful as this notion has proved to be for understanding the modernization of Western societies, the use of these terms has also generated a number of necessary critiques. One important criticism is that oppositions such as domestic and public naturalize categories and values that in fact derive from nineteenth-century Western experiences of capitalism.

The gendered spatial distinction between "inner" and "outer" in China may seem to parallel the Western categories of domestic and public, but in fact the Chinese categories of "inner" and "outer" are not so much distinct, exclusive domains as boundaries shifting along a continuum. Lan Dingyuan's statement that the roots of correct government lie in the women's quarters expresses this clearly. The fruits of *nügong*, the products of the inner chambers, integrated women into networks that connected inner and outer worlds. However, between Song and Qing times the nature of these products and of the links they created changed, and in consequence we observe a growing divergence between the social meaning that *nügong* held for the educated elite and for ordinary people.

The social value of textile production can never be reduced to its monetary equivalent, for cloth is far more than a protective layer between the skin and the air: it is a fabric, a tissue, a symbolically charged intertwining of warp and weft. Annette Weiner argues that in Oceanic societies the cloth women produced should be analyzed as a "soft" currency or possession that contributed as much to the reproduction of a society as did the "hard" currencies or possessions (precious metals, shells, weapons) produced and exchanged by men. 'In Oceania, the development of ranking and hierarchy depends upon the work of women in their economic roles as the producers of wealth and, most important, in the power of their sacredness in confirming historical and cosmological authentication.'[45]

45 Weiner (1992: 153).

Weiner's study is a critique of male-centred theories of exchange. She affirms that if analysis focuses only on male, "hard" objects of exchange, a fundamental dimension of meaning is lost. Societies are all threatened by the loss of authenticity, authority and continuity through the death of individuals, through change, through the inexorable passage of time. One way to counteract such loss is to create and pass on what Weiner calls 'inalienable possessions', precious artefacts such as family heirlooms or royal regalia. Even when they are given away, the existence of such objects, charged with symbolic meaning, serves to maintain the identity and status of the givers, 'reproducing relationships and possessions in the face of loss'.[46]

The "inalienable possessions" of Oceania were the feather capes and special cloths woven by elite women as royal and noble regalia. It is no coincidence, argues Weiner, that they are fabrics:

> Cloth may be the most apt metaphor to visualize the paradox of keeping-while-giving as societies in all parts of the world associate weaving with acts of tying and unravelling, sacred threads and dangerous dyes, woven warps and unworked woofs, expressions of longed for unity juxtaposed against the realities of death, destruction, and change.

In a society with a gendered division of labour that attributes cloth-making to women, 'intricate symbolic meanings semantically encode sexuality, biological reproduction, and nurturance so that such possessions, as they are exchanged between people, act as the material agents in the reproduction of social relations'.[47]

In the case of China, women's work of spinning and weaving was represented prosaically, in a totally demystified mode, in most of the historical documents available to us. There were no sacred texts of the weaving profession as there were for carpentry.[48] We do not know what cosmological lore or imagery of descent and renewal might have figured in women's weaving songs, for the songs are not recorded.[49] Yet the importance of exchanges of cloth between subjects and state and during the ceremonies of marriages and funerals is, I think, proof enough that textiles in China played a central role in maintaining unity in the face of 'death, destruction and change'. Quite apart from their tax contributions, women in China classically contributed three indispensable items to the ritual reproduction of community and of family: offspring, food, and cloth. Marriages could not take place without the exchange of textiles, nor could funerals be conducted without the special archaic homespun that was worn at funerals and during the mourning period. These ceremonies served to reproduce male-centred Confucian patrilines,

46 Weiner (1992: 153).

47 Weiner (1992: 154).

48 See Chapters 1 and 4, and Ruitenbeek (1993), on the *Lu Ban jing*.

49 One contemporary exception is Elizabeth Johnson's study of Hakka women's weaving of ornamental bands for their clothes (Johnson, 1977).

yet as Weiner remarks of Oceania, men depended on women to provide the "soft treasures" that constituted the key paraphernalia of these rituals.

The wife's contribution to rituals did not disappear with the late imperial changes in her productive role, but it was impoverished. After women gave up weaving they were still regarded as the people in charge of textiles within the household, so the effects of change in the textile industry in this domain should not be exaggerated. But the actual making of the cloth surely must have imparted a specifically female meaning into the rituals that was partly or wholly lost when commercial products were substituted.

Beyond the respectable Confucian interpretations of ritual textile uses, inscribed in liturgies and books of etiquette, there must have been other dimensions, magical for example, that were not recorded. To read cloth in patrilineal terms, the long threads of the warp resembled the male lineage, the weft threads wove in and out like brides[50] – but how did men view the power that lay in women's hands as the weavers of these social metaphors? And did women read these symbols of social continuity in other than purely patrilineal terms? What did it mean to a bride, when the fabrics and clothes she took with her to the new family upon which her parents had bestowed her were fabrics she had woven herself, or embroidered in a group with her mother and sisters – a pretty twist on keeping-while-giving? One suggestive ethnographic example of a use of cloth to bridge the generations and transmute loss into keeping is cited by Emily Martin in an essay exploring possible differences between female and male ideologies of death:

> If women as a gender are deeply involved in the nitty-gritty of the actualities of death, they seem in other ways to bring into funerals concerns that at first glance have no place there. James Watson . . . records that women of childbearing age, daughters and daughters-in-law, wear a piece of green cloth on their person throughout the funeral rites, in which they deal closely with the corpse and take its pollution on themselves. Later this cloth is passed over a fire in purification, and then *made into the centerpiece of the back-strap harnesses they use for carrying infants*.[51]

Even though domestic weaving declined, gifts of cloth remained an essential element of betrothal exchanges. Most of the dowry goods a bride took to her new home were cloth items of some kind – clothes or bedding as well as bolts of cloth – many of which she had woven or embroidered herself; it would be interesting to know what importance was accorded to whether certain items of the dowry were purchased or home-made. The centrepieces of the dowry were all destined

50 The warp and weft metaphor for agnates and affines occurred frequently in late imperial debates about kinship.

51 Martin (1988: 172, original emphasis), citing Watson (1982: 164–180).

for the conjugal bed, and the motifs that a young woman embroidered on her quilts, pillowcases and bed-hangings were symbols of conjugal happiness (phoenixes, mandarin ducks, and butterflies) as well as of fertility.

Although they embodied the orthodox wifely role in so many ways, nevertheless dowry goods constituted a female form of personal property that stood outside orthodox Confucian beliefs about joint family ownership.[52] As domestic weaving declined, embroidery seems to have played an increasing role in adding both economic and personal value to the goods that a bride took with her as dowry. Whether woven or embroidered, I think we can assume that these goods were particularly precious because they constituted material links with the bride's natal home, and with her mother, sisters, and other female relatives in whose company she had worked to make them. As well as a female, non-hegemonic form of property, the quilts and robes and shoes and handkerchiefs also embodied a non-hegemonic network of female kinship that was complementary to the patriline, a bond between female kin that endured beyond marriage. The embroidered goods were a medium through which the bride and her female relatives were made known to the groom's family.[53] And they were precious mementos linking women who might never meet again.

Commoners produced the basic material goods necessary for human life, and within this sphere weaving was the female counterpart of tilling the soil. The elite worked with culture, *wen*, and within this sphere embroidery was the female counterpart of writing. At the age when boys were sent off to the school-room to start grappling with the writing brush, little girls were given their first training in plying the needle. The skills were not simple: 'young girls began practising at the age of ten so they could turn out elegant pillow-cases by the age of sixteen'.[54] For fine work, one needed fingers so smooth they would not snag the silk, so rough work and fine embroidery were incompatible; young commoner women did embroider their dowry goods, but such work could not be compared to the works of art that gentry ladies created. In fact Dorothy Ko argues that in seventeenth-century Jiangnan, embroidery took on the characteristics of the male art of painting: it became a means of expressive individual creativity. Although the most famous exponents of this art were women, some male scholars also took up embroidery at the end of the Ming and in the early Qing.[55]

52 See Chapter 4 and Bray (1997: Chapters 3, 4).

53 In Ming times a bride-to-be in certain regions was expected to embroider shoes for her mother-in-law or for other female in-laws (Ko, 1994: 92, 170).

54 Mann (1992b: 262). Men also worked in commercial embroidery shops, at least by the latter half of the nineteenth century, when Western observers were often surprised at the proportion of male workers. However, the majority still seem to have been women and young girls (Wilson, 1986: 106).

55 The most famous embroiderers were the women of the Gu family, who founded what we might think of as a school of art embroidery (Ko, 1994: 173–175). In fact it was a man, Imperial Candidate Gu Mingshi, who first won renown for his embroideries in the 1550s. But the wife of his second grandson, a woman named Han Ximeng, brought the Gu embroidery art to its acme, 'blend[ing]

Embroidered clothing seems to have come into fashion in the late Ming. At the end of the seventeenth century Ye Mengzhu noted that there had been a shift from woven to embroidered designs. There is not sufficient evidence to say whether most of the embroidered sleeve-bands and slippers that elegant ladies wore were embroidered domestically or professionally, though Li Yu talks of hiring poor women as embroiderers (see below), and it was possible for women of gentle birth to earn money by their embroidery skills if necessary – as shown by the late eighteenth-century case where 'Shen Fu's future wife Yun took in embroidery work to pay her brother's tuition fees'. By the later nineteenth century the Western appetite for Chinese embroideries had gentlewomen in towns such as Ningbo hard at work, and Lucy Soothill observed that in poor families women and girls were sometimes able to help out by working at embroidery.[56]

In the Ming and early Qing, embroidery had a rather ambivalent status in the eyes of elite men. For an agrarian like Zhang Lüxiang, writing of how to make a decent living by farming, embroidery was a frivolity that distracted from proper work – women should weave cloth, not decorate it. For well-off families it was the symbol of the leisure they could give their women: it marked them off from the rough-fingered peasant women toiling at rush mats, and it contributed symbolic rather than material capital, an enhancement to the family's material surroundings produced within the seclusion of the inner quarters by women whose embroidery skills marked them as gently nurtured. Because there was as yet no real commercial market for fine embroidery, it was work and yet it was not work, 'the hallmark of Qing domesticity'.[57]

Both Lü Kun and Chen Hongmou deplored the fact that parents who loved their daughters were satisfied with training them to embroider. This they saw as second-best, an inadequate moral and practical education: instead girls should be taught to read so they could study moral principles for themselves. Li Yu on the other hand complained of the fact that gentlewomen of his time were abandoning embroidery for letters: 'they pursue only men's skills, and sneer at women's work . . . [and] even hire old women and poor daughters to make their three-inch arch shoes for them'.[58]

As far as a woman's view of work and kinship was concerned, embroidery was very much productive labour in the social and symbolic sense, "womanly work". The rituals of the Double Seventh festival – needle-threading competitions and

the theory of painting with the skills of embroidery to the betterment of both' (Gao, 1986: 34). Several of her works have survived, and most of them bear her seal – the mark of art, not of craft (Huang, 2012). On the spiritual dimensions of certain styles of embroidery, see Li (2012).

56 For Ye Mengzhu, see Wilson (1986: 58), and for the case of Shen Fu's bride Yun, see Wilson (1986: 105). On the Ningbo embroidery industry, see Mann (1992b); Wilson (1986: 105) mentions Lucy Soothill's comments.

57 Mann (1992b: 260).

58 Handlin (1975: 36–37); Ko (1994: 175).

the like – were celebrated by women throughout society. Girls in every class would leave their families behind on marriage and go to live with strangers, and embroideries accompanied them on this frightening journey. Even an ordinary girl embroidered the shoes and quilts she took as dowry to her new home; the hard work and skill she had devoted to the task showed her fitness to serve as a daughter-in-law. Few women were literate, but embroideries or woven patterns conveyed messages of affection and solidarity that otherwise could not have been expressed, and affirmed bonds between related and unrelated women that neo-Confucian orthodoxy did not encourage, or at any rate hardly recognized. Where girls formed "sworn sisterhoods", they would sit together in the evenings in their girls' house embroidering and telling stories.[59] The objects that women exchanged as tokens of friendship were mostly embroidered – for instance, handkerchiefs, fans, or shoes.[60] Whereas textiles woven for tax payments or commercial sale linked women into broader networks of governance and livelihood, embroideries symbolized an affective life that reached outside the walls of the inner quarters, a female network that transcended the spatial boundaries laid down for women by their male kin.

Women's work and patriarchy

Henrietta Moore remarks that:

> cultural ideas about gender do not directly reflect the social and economic positions of women and men, although it is true that they originate within the context of these conditions. This is because gender stereotypes are developed and used in the strategies which individuals of both sexes employ to advance their interests in various social contexts.[61]

This stricture applies not only to the relations between ideology and material conditions within the societies that we study, but also of course to our scholarly analyses of other societies. In order to explore the meanings of women's work in imperial China, not only do we have to try to disentangle the relations between representations and experiences of the time, we also have to recognize and account for the stereotypes we have created for ourselves. I think this is especially important when we try to understand how work and changes in work fit into perceptions of the relations between men and women in late imperial China.

59 Topley (1978); Stockard (1989).

60 Poems and other texts written in the famous "women's script", *nüshu*, of Hunan were often written on pieces of embroidered fabric (see Silber, 1994, and the plates in Gao and Yi, 1991). Ko (1994: 170) gives examples of poems written in thanks for a gift of embroidered shoes.

61 Moore (1988: 37).

In the retrospective analysis of female roles in pre-modern China, reproduction has loomed much larger than production. Until recently, little theoretical attention was paid to the place of productive labour in the construction of pre-modern women's roles. Daughters were represented as essentially no more than future brides, and wifely identity was construed principally in terms of biological motherhood. Though at one level this characterization of the wifely role is an accurate reflection of late imperial Chinese sensibilities, one cannot help observing how closely the womb-focused image of Chinese wives mirrors our own cultural predilection for construing female identity in terms of biological reproduction and its control. Moreover, as Dorothy Ko and others have pointed out, such constructions of "traditional" women in China are rooted in distorted and totalizing images generated by the nineteenth-century confrontation between China and the Western powers. In the late nineteenth and early twentieth centuries, Western writers as well as Chinese critics of the crumbling Qing state elaborated an orientalist image of "traditional" Chinese woman – imprisoned in the inner quarters, deprived of all freedom and true dignity, a powerless object of exchange between patrilines, a baby-machine physically deformed by the practice of foot-binding and morally deformed by the tyrannies of patriarchy, so crippled she was incapable of productive work – an image that still directs the rhetoric and scholarship of many politicians and historians, not to mention the general public, today. We have to bear in mind that this feminine image was not so much a true portrayal of the condition of Chinese women at the time as an expression of the historical crisis that then faced the Chinese nation.

One reason why Western perceptions of the wifely role in China have continued for so long to be preponderantly negative relates, I think, to the distaste that modern Western feminists feel for distinct and exclusive gender roles (especially if associated with social segregation of the sexes). The fact of women's seclusion in imperial China has disposed scholars to visualize women's work as being restricted simply because it took place within the inner quarters; we have therefore assumed that it was more or less limited to the reproductive activities that our own historical experience of industrialization leads us to associate with the domestic sphere. In the nineteenth and twentieth centuries, foreigners and Chinese reformers argued that a nation that bound the feet of its women and deprived them of education was crippling itself by crippling half its subjects. Bound feet supposedly confined women to the inner quarters and excluded them from productive labour. However, this was a double misunderstanding; first of what female seclusion in China was about, and second of how productive labour was organized in a pre-industrial economy.[62]

62 And also of the extent to which bound feet impaired women's mobility and their capacity to work. See Blake (1994) and Mann (1992b) on women's important contributions to the home-based industrial economy of rural and urban regions in the late Qing; Blake (1994) and Huang (1990: 55–57) discuss women's contribution to farm work outside the home.

Nevertheless, the essentialization of "traditional Chinese woman" certainly found some basis in the realities of the period during which it was formulated: late Qing women were secluded and subject to many forms of male control, they did have bound feet, and – as I have argued – their role as economic and political actors was in many cases much less visible to themselves and to those around them than it had been in earlier times.

The historical changes in the textile industry in late imperial China served, I believe, to consolidate a reconfiguration of gender roles structured by changes in kinship practice and in inheritance and property law. They provided fertile ground for popular acceptance of doctrines on male dominance and female subordination that were widely circulated during the late Ming and Qing, whether in the form of printed primers on wifely virtues, of lectures upholding Confucian values given in ancestral halls or before the magistrate's office, or of the sections on family ritual in almanacs and popular encyclopedias.

As the nature of the female contribution to the family economy changed, its increasing invisibility (in all but a few cases) supported a revised representation of the wifely role as complementary but dependent, concerned primarily with reproductive rather than productive tasks. I do not mean by this that all Chinese women became victims of their own wombs. Elsewhere I argue that in late imperial China reproductive tasks were not necessarily reducible to child-bearing, but were also recognized as including household or even estate or business management, and the raising – as opposed to the bearing – of children.[63] The pressures on an individual woman to prove her worth by bearing children depended on the extent to which her class and status allowed her alternative expressions of her womanliness. In elite families, I argue, the women at the top of the household hierarchy were valued as educators and managers, and were able to exploit the fertility of women of lesser status. In working families, I agree with Gates that the conditions that emerged in China from Song times allowed the patriline to appropriate the products of women's labour increasingly effectively, while the decline in the perceived value of female productive labour helped consolidate an image of women as subordinate and dependent. Except where some form of remunerative "women's work" was available, increasingly it was felt that the most valuable good a woman could contribute to her household was offspring.

63 Bray (1997: Chapters 7–9).

4

STRUCTURES OF FEELING

Decorum, desire and a place of one's own

Every society evolves characteristic conceptions and uses of space, and as a society changes so too do its spaces. As Henri Lefebvre, probably the most influential theorist of space, insists, this relationship is dialectical: a change in the concept of space is not simply the expression of social change, it is also its vehicle and its agent, 'its tomb as well as its cradle'.[1] Each mode of spatial experience, moreover, embodies particular apprehensions and layerings of time – whether it be the comforting nesting of personal rhythms of birth, marriage and death within the higher passions of the annual liturgical cycle inside the portals of a Christian church, or the abstracted space-time of modern science, where both dimensions of being lose their sacred, aesthetic and moral significance as they are first represented, then experienced, as neutral, uniform, linear modes of extension, 'the organizing medium of modernity's institutional dynamism'.[2]

In the case of modernity as it emerged in the industrializing West, we have long recognized that transformations in the experience of space and time went hand in hand with the development of new relations of production, new economic and political networks, and new forms of social, civic and psychological being. In analyzing the nature of various modern apprehensions of time, of space and of the relation between them, and in tracing the intellectual, aesthetic and social processes through which corresponding forms of perception and spatial practice emerged, historians and social theorists have greatly enriched our understanding of what it means to be modern, and in doing so they have created powerful tools for

1 Lefebvre (1991: 31, 34).

2 Giddens (1984: xii).

examining the nature of other, earlier societies. In this chapter, I draw on insights that the theorizing of modern space has brought to our understanding of modern being, to explore the kind of space or spaces that late imperial Chinese society produced for itself and to consider how these spaces accommodated the expression of orthodoxy or of human desires.

I concentrate on late imperial space in its domestic forms because, as I argued in Chapter 1, the house and its spaces were of paramount importance in reproducing orthodoxy and shaping subjectivities in late imperial society. The focus on domestic space is also rewarding because there are several ostensible parallels between patterns of change in China and the modernizing West. These, I feel, deserve attention given the key role that new representations of domestic space are considered to have played in the formation of modern Western identity, desires and gender roles. How do we explain the paradox, for instance, that in advanced economies such as Britain or the United States until just a few decades ago the confinement of respectable women to the domestic sphere was considered a sign of advanced civilization and high status, a goal to which upwardly mobile families aspired – while a suspiciously similar separation of the sexes in "traditional" China (or elsewhere) was considered primitive, oppressive and demeaning?

First, then, I consider the roles attributed to domestic space in the shaping of modern Western subjectivity and review the complex approaches to spatial meaning that have come out of our attempts to theorize modernity. Next I consider four representations of domestic space-time in late imperial China to ask what they reveal about identities, orthodoxy and desires in that society, and in particular about different registers within which gender differences could be experienced. I conclude with a brief discussion of what this kind of comparative history suggests for linking spatial change, historical change, and passages to modernity.

The house and modern being: contending architectonics

One often noted feature of modernity is its "emptying" and separation of the categories of space and time, their disenchantment. Trimmed of social, sacred, aesthetic or moral significance, they become universal units that facilitate efficient and effective action, equally useful to governments and to business. Such rationalization of space and time is, in Marxist terms, both alienating and mystifying. In order to expose and transcend the limitations of this reductionist notion of modern or capitalist space and reach a more accurate understanding of space in the production of social worlds and of ideology, French intellectuals developed the concept of *l'espace social*, social space. This conceptualization aims to give texture, structure and definition to our spatial experiences and representations, to re-connect space and time, to saturate the "empty" space of science with human bodily presence, perception, action and power. Social space is at once a field and a basis of action, offering its occupants both concrete and conceptual resources.[3] *Architectonics* is the term proposed by the French Marxist philosopher Henri Lefebvre to define the task of disentangling the layers of time and meaning embedded in social space:

> In space, what came earlier continues to underpin what follows. The preconditions of social space have their own particular way of enduring and remaining actual within that space . . . The task of *architectonics* is to describe, analyze and explain this persistence, which is often evoked in the metaphorical shorthand of strata, periods, sedimentary layers and so on.[4]

Borrowing Heidegger's phenomenological term, we might say that architectonics investigates and relates different modes of being-in-space, or – for the purpose of this essay – of *being-in-the-house*.

Homogeneous, neutral time-space is certainly a ubiquitous feature of the modern world. Science, technology and law have given it power and claims to universal validity, common sense accepts its usefulness. Yet at the same time the critics of modernity have made it clear that what we identify as the quintessential or even the only modern vision of time-space in fact coexists with alternative apprehensions of space and time that are often mistakenly treated as pockets of un-modernity, "traditional" survivals or ineradicable atavistic yearnings. In fact, the obviously modern and what we might best call the anti-modern (since this terms implies a dialectical relation rather than the distinctness that a term such as traditional implies) are inseparably entwined and interdependent. However traditional and un-modern these other visions of space and time may seem, they are an integral element of modernity, a variegated and constantly evolving reworking of the past (or future) to serve the present that tells us as much about ways of being modern as the study of the industrial assembly line.

In the modern West, a house is thought of as a *home*, a place of privacy and of family, a haven in a heartless world, a repository of memory and a refuge from the alienation, exploitation and selfish competition of the marketplace. We value our houses as an expression of our deepest selves. We require them to provide experiences that are non-modern or perhaps better anti-modern, they are places where we hope still to experience space and time as entwined and infused with meaning.[5] Nostalgia is the natural daughter of modernity. If we as moderns see

3 Lefebvre (1991: 191).

4 Lefebvre (1991: 229).

5 In the United States, the most mobile of modern societies, there is little opportunity for the majority of the population to infuse their living spaces with shared memories for they move on average thirteen or fourteen times in a lifetime. Perhaps that is why in the United States houses are always sold as "homes", implying that the four walls come not only with inbuilt closets but also with inbuilt emotional warmth.

our public, economic lives playing out in the neutral, forward-looking space and time of science and commerce, we need to imagine our "home" as a place where richer, more traditional rhythms of space, place and emotion obtain. Gaston Bachelard elaborates this visceral craving for an alternative to the arid space-time of science in his essay *The Poetics of Space*, drawing attention to the oneiric dimensions of intimate spaces, to the emotional comfort offered by drawers and cushions and their affinities with natural analogues like nests or shells, and to the strength of our urge to identify 'Home and Ego'.[6]

Our current expectations of home are the outcome of a long historical process of industrialization, new divisions of labour, changes in property law, the growth of consumerism and new moral expectations of men and women. Historians have richly documented how new models and norms of domestic space helped build modern identities and subjectivities, whether in England or New England, France, Japan or the Soviet Union.[7] In the course of the nineteenth century the very structure of the house in industrializing societies and their colonies became a "normalizing device".[8]

In England, for instance, the bourgeois concept of home and the cult of domesticity began to crystallize in around 1820. The concepts of house, home and family were merged and represented, first by evangelical preachers then by mainstream moralists, as a place of piety and moral development both for family members and servants. The nature of home was defined in distinction to the outside world, and in particular to the savage Hobbesian world of the marketplace where men laboured to support their families, or to the political arena of clubs, pubs, meeting-houses and parliament. The public arena was the male domain, home was the domain of female qualities and values. Women were responsible for home and family but should not have to work for money; their role lay in nurturing the family and providing a physical and moral haven where men could regain strength. '"The sweet caresses and endearments of wife and children" were to be set apart from "the cares and anxieties of business"'.[9]

This sweet domesticity required a redesign of living space. New "dedicated spaces" appeared that corresponded to emerging conceptions of identity and sociality. The distinctness of childhood was embodied in the physical separation from adult quarters of the nursery and the school-room, the drawing room provided a setting for genteel mingling of gentlemen with ladies, and table-manners in the servants' hall mirrored the hierarchies and etiquette of bourgeois

6 Lefebvre (1991: 121), commenting on Bachelard (1957).
7 See, for example, Perrot (1990) on Europe, Hayden (1986) on the United States and Sand (2004) on Japan.
8 See Chapter 1 and Bray (1998).
9 Hall (1990: 71).

society. The family house, standing alone in its gardens, where masters, children and servants assembled each morning for prayer then went their separate ways, became the desirable norm for the English middle class. In 1851 the category of "housewife" was added to the English census.[10]

Social reformers pressed this ideal of domesticity, together with the ideal of home ownership, on the working class with considerable success. Unions and other working-class groups established building societies and other savings and loan institutions to allow their members to achieve the dignity and comfort of their own house. Capitalist employers were not blind to the fact that a man with a mortgage to pay each month is less likely to go on strike than a man with fewer financial commitments. Co-opting the lower classes through consumption that raises their social status and their stake in the system has been one of capitalism's most effective strategies; increases in wage bills are offset by the corresponding expansion of markets for consumer goods and also, usually, by a drop in revolutionary fervour. Since the mid-nineteenth century, expanding both the numbers and levels of middle-class domestic consumption has been one of the economic driving forces of Western capitalism. Middle-class lifestyles depend on the possession of certain goods, and by acquiring the goods one may hope to enter the middle-class, which thus absorbs ever-greater sections of the population. It was "selling Mrs Consumer" cars, refrigerators and television sets, that more than any other economic venture, helped restore the shattered economies of the industrial West after the Depression and World War II.[11]

The emerging consumption patterns of the nineteenth-century middle classes naturalized the standards of living and relation to our bodies that we now think of as unproblematic universal needs. In his magnificent study of technology, values, and modern being, *Mechanization Takes Command*, the architect Siegfried Giedion (1948) traces the origins of modern comfort, with its emphasis on adapting the environment to the body rather than the body to the environment, and its naturalizing of posture and motion by erasing their social or moral content.[12] For example, anatomical representations of the body as a machine, together with the detailed study of movement for which photography paved the way, led to the development of the science of ergonomics, which was swiftly applied to the development of furniture that allowed the body to work or rest more efficiently. Meanwhile, ideas about germs, dirt and contagion emerged that overcame a deep, society-wide and long-standing Northern European antagonism to washing and popularized the ideal and – for the better-off – the reality of daily baths, frequent laundering of clothes (necessitating of course the purchase of *more* clothes) and the

10 Hall (1990: 74).

11 See Mandel (1975), Cowan (1983) and Hayden (1986).

12 See also Rybczynski (1986).

conquest of household dirt.[13] Contraptions were patented in profusion for washing and mangling laundry, for beating carpets, for heating water conveniently and cheaply. The tin bath, carried into the bedroom when needed, was replaced by a dedicated *bathroom* filled with bulging pipes, gleaming enamel and fluffy towels. A flush toilet sat next door in "the smallest room", sparing maid and master from what had now become the embarrassment of the chamber pot.

In the Introduction, I presented the flush toilet as an icon of the modern, the embodiment par excellence of the particular forms of cleanliness and privacy that are essential to the psychological well-being of modern Westerners. Privacy is a desire, a form of being-in-the-world that is recognized as fundamental to the constitution of Western modernity, underpinning the notion of the individual, private property rights, individual consciousness and so on. The modern house or home is valued as the locus of privacy, distinct from the public domain – not only from the public domain of work but also from that of politics and the state. Although, in fact, modern states penetrate deep into every form of family activity, while evolving family ideologies shape the nature and functions of the state,[14] one of the key beliefs of modernity is that we let the world into our homes only on our own terms. So we see the toilet as quintessentially private, the least public place of our life, whereas in fact its use depends on a vast technical infrastructure funded and maintained by the public sector.

The separated spaces of the modern home offer scope to "be oneself", giving shape to new desires for intellectual and psychological satisfaction – think, for example, of Virginia Woolf's choice of "a room of one's own" as the metonym for female artistic emancipation, or the contemporary North American middle-class belief that every child needs her own bedroom, bathroom and phone if she is to thrive as a teenager. The domestic architecture of the modern-self emerged in response to the new-forming desires of the nineteenth-century bourgeoisie, and this dominant class was not slow to see the possibility of applying household design to the improvement of the lower orders. Whereas seventeenth- and eighteenth-century planning had concentrated on designing ideal towns and left interiors up to the occupants, in the nineteenth century town planning went hand-in-hand with the design of house layouts that would provide morally and physically healthy environments for the working class. In the Manchester Museum of Science and Industry there is a dramatic reconstruction of a typical industrial workers' dwelling circa 1800: it consists of a windowless hovel a few feet across, in which children share the muddy floor with the family pig. The endless rows of terraced houses that paternalist industrialists were providing for their workers by mid-century offered

13 On the revolution in standards of cleanliness, see, for example, Giedion (1948), Corbin (1982) and Cowan (1983).

14 Robertson (1991).

what they considered spaces of respectability and hygiene: separate bedrooms for parents and children reduced the risk of promiscuity with its social-Darwinist association of degeneracy; the shared privy and water tap in the yard allowed families to achieve cleanliness and reduced the risk of contagion. On the one hand, then, the spatial design of the back-to-backs disciplined the worker families and co-opted them into the value-system of the bourgeoisie; on the other, this social space was appropriated (in de Certeau's sense) by working-class families to construct an alternative to capitalist exigencies, a network of family support and a sense of community, whose value was thrown into relief when national prosperity and desires to diffuse higher living standards led governments on both sides of the Atlantic to initiate slum-clearance programmes and resettlement into high-rises in the 1960s and 1970s.[15]

Lefebvre labels professional conceptions of how space should be designed and produced *representations of space*. In the modern experience, science, says Lefebvre, has defined the dominant representations of space, and through applications such as domestic science and the scientific design of dwellings has shaped a modern sense of self in which capitalist desires are absorbed through everyday spatial practice by all members of society. 'The spatial practice of a society secretes that society's space; it propounds and presupposes it, in a dialectical interaction.'[16] Against the moralized spaces of ordered docility proferred to their labour force by the capitalist bourgeoisie, socialist planners in the Soviet Union, Scandinavia, Bauhaus Germany and the Israel of the kibbutz movement devised what they viewed as emancipatory domestic spaces for workers, spaces that tempered the virtues of privacy and independence with the advantages of sharing. Many such designs went beyond streamlining household chores to reducing loneliness and isolation for small children, old people, and especially women: common laundry, catering and even childcare facilities were designed to free the working-class housewife from drudgery and social isolation, to free her indeed to join the waged workforce.[17]

The scientific representation of domestic space inculcates a yearning for efficiency and convenience, met principally by substituting machines or processed goods for human limbs and raw materials: vacuum cleaners were advertised as replacing flocks of maids, washing machines washed whiter than bare hands, sliced bread protected children from the horrid wounds that unskilled wielding of the bread knife might inflict. Bachelard's dreamspaces might seem to embed non-capitalist desires, yet it is no paradox that this capitalist-engendered desire for a private world beyond the marketplace is itself highly marketable. Astute entrepreneurs such as Ralph Lauren have cashed in on our nostalgia, twin sister

15 See, for example, Roberts (1971) on Salford, or de Certeau (1984) on Croix-Rousse in Lyon.
16 Lefebvre (1991: 38).
17 Hayden (1986).

of our belief in the future, by offering us a choice of ready-made traditions to wrap around us in the form of coordinated sofa covers, curtains and sheets.

Like the house itself, these "traditional" styles offer a dual possibility of self-fashioning through distinct spatial representations. One popular form offers association-by-possession with an imagined aristocracy, feeding the desire for upward mobility that Elias and Veblen have dissected so skilfully as one of the most powerful motors of capitalist development from *ancien régime* times on. Another feeds the modern yearning for nature explored by Raymond Williams in *The Country and the City*, our desire to taste the earthy aesthetics and authentic tastes of an invented peasantry that we locate not in muddy hovels but in picturesque stone villas, preferably in Tuscany.[18]

Theorizations of modern space

In Lefebvre's terms, then, modern middle-class domestic space contains multiple overlapping *representational spaces*, that is to say 'space as directly *lived* through its associated images and symbols'.[19] For Lefebvre space is a social construct, both a product and a producer of social relations, that has to be understood as belonging at once to a material base and to an ideological superstructure. He is concerned with how perceptions and intellectual representations of space relate to power and to historical change. As a Marxist writing in the 1970s, Lefebvre is especially interested in the history of the production of late capitalist space, which he qualifies as abstract and alienating, a 'phallo-video-geometric' space, the roots of which he traces back *inter alia* to late medieval urbanization in Italy and the development of perspective in Italian painting; to Cartesian homogeneous geometric space; and to the progressive twentieth-century substitution of space for land as a commodity and factor of production.[20] Individuals within a society have different class interests and thus tend to prefer certain available representations of space over others, yet in the course of their daily lives they will continually encounter and assimilate alternative representations. Like Raymond Williams writing on the culture of technology, Lefebvre seeks to break open the concept of hegemony to clarify the interaction between residual, dominant and new representations within a particular society. In modernist spirit he treats the production of social space as a historical *process*.

While Marxists such as Lefebvre, Williams and Harvey believe that it is possible to trace the emergence of new conceptions of space historically as part of the

18 Veblen (1934 [1899]); Elias (1985 [1933], 1982 [1939]); Williams (1973).

19 Lefebvre (1991: 39).

20 Themes further developed by David Harvey in his analysis of the role of space in late-capitalist accumulation (Harvey, 1990).

constituting process of the modern, thinkers such as Bourdieu and Giddens are perhaps more post-modern than modern in resisting epistemologies of explanation and causality, resorting instead to more diffuse notions of predisposition or sedimentation, that by their nature avoid linearity. However, unlike Bachelard, who proposes a *poetics* of space rather than an alternative rationality, Bourdieu follows in the semiotic footsteps of Durkheim in proposing structured or at least structuring principles by which space is differentiated.[21]

Bourdieu developed the concept of *habitus* to get beyond the timelessness, the neglect of individual agency, of active choice between strategies, of history that he saw in both functionalist and structuralist anthropological treatments of culture. Rejecting the use of terms such as "norms" or "rules", Bourdieu instead invokes *habitus*, a generative and structuring principle of strategies and social practices, a system of dispositions.[22] As these principles generate practices, they reproduce the conditions which gave rise to them initially – so habitus can be understood, according to Bourdieu, both as product and producer of history; furthermore, he believes that it bridges the gap between individual behaviour and social change.

Bourdieu sees the house as a principal site for inculcating habitus, which often consists of tacit rather than explicit ideas and knowledge – as Foucault realized, this is often the most powerful way to believe. A child acquires this kind of knowledge about, for example, symbolic oppositions, not through verbal instruction but by imitating the behaviour of her elders. For Bourdieu the house teaches tacit knowledge, what Giddens calls practical as opposed to discursive consciousness.[23] The power of embodied knowledge is all the greater because it is unconscious and therefore unexamined.

'Habitus', says Bourdieu, '[is] embodied history, internalized as second nature and so forgotten as history.'[24] Although Bourdieu also claimed that habitus offered a tool for theorizing history in the making, in fact he turned the concept to best account in studying the reproduction of class distinctions.[25] Norbert Elias, on the other hand, developed a rather different concept of habitus intended explicitly as as a tool for dynamic historical analysis – analysis that would account for processes of change (what he called socio- and psychogenesis) as well as reproduction. Elias defined habitus as the "psychic economy" characteristic of a specific social group or "figuration" and remodelled through shifts in the balance of tensions between figurations. Elias felt Marx was mistaken in believing that such dialectics were essentially dualistic, played out in the conflict between dominant and dominated

21 See especially Durkheim (1915: 11).

22 Bourdieu (1976: 214). Bourdieu's classic analysis of domestic space as habitus is his essay on the Berber house (Bourdieu, 1973).

23 Giddens (1984: 7).

24 Bourdieu (1990: 56).

25 Bourdieu (1984).

class; he also held that the economic dimension is only one of several that fundamentally shape the relations between classes and the identities of persons belonging to them, so that "ideology" is not superstructural but far more central.[26]

Elias, like Bourdieu, believed that one of the most powerful forms of the expression and experience of ideology – as well as the most easily retrievable – was its concrete, material manifestations. For that reason he begins his study of *ancien régime* France, *Court Society*, with a chapter entitled 'Structures and significance of the habitat'. In the Introduction, I introduce Elias' analysis of a new form of building characteristic of this period, the *hôtel* that was the Paris residence of the grand aristocratic family. Elias explores not only the values displayed and the social relations expressed by the spatial design of the *hôtel*, but also the meaning of the modifications that the bourgeoisie made to this style of dwelling. He stresses that in a society where the political players fell into three groups, the absolute monarch himself, his courtiers and the aspiring bourgeoisie, the aristocratic *hôtel* only made sense in relation to Versailles on the one hand, and to the bourgeois dwelling on the other.[27]

Beyond group or class dialectics as analyzed by Elias, we also have to consider how historical changes in social space reflect adjustments of emphasis or shifts in the balance of power between coexisting representations of space and representational spaces within a group or class. A house should not be viewed as straightforward symbolic system: the anthropologist Ellen warns of the dangers of an overly rigid and static structuralist view in which 'the symbolic order is somehow frozen into the fabric', and draws attention to interpenetrating levels of meaning of different kinds of symbolic order that are brought together in the house, noting the possibility of contradictory expressions of symbolic meaning, and stressing the house's 'multiple significance' as an organic whole.[28] Or, in the words of Michel de Certeau, we can view a space as a 'polyvalent unity of conflictual programs or contractual proximities'.[29]

De Certeau makes a useful (if somewhat confusing) distinction between space/*espace* and place/*lieu*, a distinction that is analogous to that between speech/*parole*

26 Elias began his research working with Karl Mannheim at the University of Frankfurt in the 1920s and 1930s. His studies first appeared in German and were not translated into French until relatively late (Elias, 1985 [1933]), so it is perhaps not surprising that Bourdieu appears to have been unaware of Elias' work on habitus.

27 For example, aristocrats were not concerned with what their servants saw or thought, an attitude that was to the bourgeois mind immoral. Whereas the aristocrats justified an action in terms of its effects within their own group, the bourgeois extended their own morality and feelings to their social inferiors. This can be seen in the much more permeable boundaries – spatial as well as behavioural – that the bourgeois drew between themselves and their servants.

28 Ellen (1986: 4); see also Carsten and Hugh-Jones (1995).

29 de Certeau (1984: 117).

and language/*langue* – that is to say, place defines order and stability, while 'space is a practised place'.[30] Orthodoxy, then, would be inscribed in *place*, while multiple versions of *space* could be played out in that same place. The variations of space might correspond to different actors, different relationships, or different interpretations of place coexisting in the mind of the same person. *Stories* transform places into spaces (maps into tours) and vice versa.[31] We can then view any *place* (let us say, a Chinese family house) as containing a multiplicity of *spaces*, as consisting of multiple architectures of meaning.

In an early trend-setting set of studies of space and gender, Shirley Ardener and her colleagues documented how social relations structure the meanings of shared space for the different persons in it. The same space can have different meanings for men and for women, and also can take on different meanings at different times (for instance, Iranian tribeswomen who will move freely and unveiled through their village during the day while the men are away with the herds leave their houses reluctantly and carefully veiled at night when the men have returned).[32] Allan Grapard reads the landscape of Tokugawa Japan as simultaneously containing multiple representations of meaning, multiple desires, multiple understandings about how knowledge is gained; he argues that the three different philosophies of space he proposes (geosophia, geognosis and geopiety) correspond to class differences, to different senses of politics and to different understandings of time.[33]

Recent theorizations of space, then, insist on the need to re-connect space and time, and to recognize the existence of concurrent and possibly contending epistemologies and desires inscribed in any single landscape or place. In what follows I shall attempt to show how time and space connect in different representations of late imperial domestic space, as well as exploring the desires and emotions to which they allow or deny expression.

An architectonics of Chinese being-in-the-house

In the classical Chinese language one single term, *jia*, refers to the physical structure of the house, to the family that inhabits it and to the sense of home or community that such occupation produces. Family relations were not only located in but also constituted by space, and personalities were entwined in the spaces and spatial relations they created and inhabited.[34] As I argue in Chapter 1, spatial forms and

30 de Certeau (1984: 117). His assignation of the terms space and place is confusing because the terminology is reversed by most writers on the issue, including myself.

31 de Certeau (1984: 118).

32 Ardener (1993); see also Moore (1986) on shifting genderings of space in response to modernization, and Abu-Lughod 1986 on the rich meanings of veiling and unveiling in Egyptian Bedouin society.

33 Grapard (1994).

34 See Knapp and Lo (2005).

practices provided a template for orthodoxy in late imperial times, yet the manipulation and practices of domestic space allowed room for the expression of other emotions and desires, sometimes simply different "styles" of "using" the orthodox template, sometimes styles that were implicitly or even frankly "delinquent"[35] – though to my mind such forms of rejection or modulation of the orthodox are often better thought of as transgressive or subversive than as forms of resistance.[36]

In Chapter 1, I explored the interplay between two "imaginary architectures" (or in Lefebvre's terms representations of space), mapping the interpretative schemes they jointly offered for creating and inhabiting domestic space in late imperial China, and their power to bind increasing numbers of families into orthodoxy while accommodating aspirations and emotions that diverged from orthodox ideals. Here I propose an elaborated, more gender-focused interpretation of those architectures of *decorum* and of *energetics*, to which I add two further categories: the architectures of *aesthetics* and of *comfort*. I develop the theme of polyvalent spaces to explore different narratives or stories of what is essentially the same physical place, that is, the house of a family in comfortable circumstances – not a top-ranking family, nor yet a poor family, but a family well enough off to occupy a house that contained all the elements required for respectable life. In the spirit of Lefebvre, I try to develop an architectonics of domestic spaces, to consider what desires, emotions and sense of time they might express, and the images and practical expressions of gender relations that they imply. I illustrate the four forms of spatial sensibility through four vignettes, four representations of what domestic space looked like, how people behaved in it and which sentiments were appropriate to it.

My earliest vignette and its corresponding "story" dates back to the eleventh and twelfth centuries, to Sima Guang's *Jujia zayi* (Miscellaneous etiquette for family life) and Zhu Xi's *Zhuzi jiali* (Master Zhu's Family rituals); these were the texts that established the neo-Confucian orthodoxies of domestic ritual, etiquette and spatial practice for centuries to come. My latest "story", traced in a Japanese compilation entitled *Shinzoku kibun* (Recorded accounts of Qing customs), dates from 1799 and consists of accounts of the house collected from seven Jiangnan merchants on the initiative of the magistrate of Nagasaki; one interesting aspect of this document is that by this point in late imperial history the merchants are confident of the respectable status of their domestic spaces. My two other stories are the *Lu Ban jing* (Carpenter's canon), a medieval carpenters' manual the earliest portions of which date back to the Song, and *Xianqing ouji* (Casual expressions of idle feeling), published in 1671 by the famous aesthete Li Yu. Here, then, I have an orthodox representation of being-in-the-house that by 1500 had diffused its morality to all

35 de Certeau (1984: 30, 130).

36 See Bray (1997: 46) for my reservations about over-enthusiastic searches for signs of resistance.

levels of society, a geomantic story about domestic space that was also common knowledge by the late Ming, and two more variations upon the same spatial theme. Each text, I shall show, represents a different set of ideals, morality and desires, and different models of gender relations.[37]

The space of decorum and the architecture of orthodoxy

As discussed in Chapter 1, the orthodox neo-Confucian representation of domestic space portrayed a house centred around the ancestral altar, a space of decorum. The spatial configuration and practices associated with this form of being-in-the-house were first systematically laid out in Zhu Xi's *Family rituals* of c. 1169, which was later to circulate in numerous editions and versions.[38] The *Family rituals* was intended to provide a comfortable liturgical handbook for everyday use by ordinary people without a high level of education. It prescribed ideal forms of ritual practice that would replace heterodox or inadequate ritual, and consolidate an orthodox morality based on the values of seniority, love and respect. Zhu Xi selected between ritual instructions for routine ancestor worship and for rites of passage, as prescribed in the works of other Song philosophers. The five chapters of the *Family rituals* are on the general principles of ritual; on initiation rituals into adulthood (capping for boys, hair-pinning for girls); on weddings; on funerals; and on sacrifices:

> One of the most important of [Z]hu [X]i's own ideas was to create a new first chapter to highlight the centrality of the offering hall and the more routine ceremonies held there, including daily "looking in", twice-monthly "visits", and more formal offerings on popular festivals like New Year.[39]

Zhu Xi's first chapter, 'General principles of ritual', begins with a discussion on the offering hall, which he explains as a space for fostering the emotions of respect and duty:

> This section originally was part of the chapter on sacrificial rites. Now I have purposely placed it here, making it the first subject, because its contents form the heart of 'repaying one's roots and returning to the beginning', the essence of 'honoring ancestors and respecting agnatic kin', the true means of preserving status responsibilities in the family, and the foundation for

37 For further details on these key texts, see Bray (1997: 59–70). Another perspective is provided by Dardess (1989), who analyzes the concentric social spaces radiating outwards from a manor house in Jiangnan in the Ming.

38 See Ebrey (1991a), whose authoritative translations I quote.

39 Ebrey (1991b: 105).

> establishing a heritage and transmitting it to later generations. My arrangement will let the reader sense that what is placed first is the most important.[40]

What we have here, then, in Lefebvre's terms is a representation of space that gives new moral resonances to lived, representational spaces and transforms the house into a disciplinary space. In de Certeau's terms, we have a story that turns space into place, the inscription of an orthodoxy of practice, emotion and desires on the everyday domestic features of the Chinese family house.

Zhu portrays a household whose life revolved around the shrine. It was the heart of the house, its most precious possession: 'Should there be flood, fire, robbers, or bandits, the offering hall is the first thing to be saved. The spirit tablets, inherited manuscripts, and then the sacrificial utensils should be moved; only afterward may the family's valuables be taken'. The shrine was something of which all members of the family were conscious at all times. The main ritual actor in this domestic space of decorum was the "presiding man", *zhuren*, the senior patrilineal heir. He was expected to look in on the shrine every morning, burning incense and bowing. His wife, the "presiding woman", *zhufu*, was his partner in all of the more elaborate ceremonies, and like him was expected to announce all her comings and goings to the ancestors. 'When the presiding man and presiding woman are about to go some place, before departing they enter the outer door of the offering hall and perform the "respectful look". They do the same on returning'.[41]

The *Family rituals* includes most of Sima Guang's *Miscellaneous etiquette for family life* at the end of the first chapter. The etiquette that Sima (1019–1086) prescribes for domestic routines is a useful complement to the liturgies prescribed by Zhu: as with other normative neo-Confucian texts on deportment and management, the observance of spatial boundaries is a central theme. The following famous passage on segregating the sexes was, the well-known moralist Yuan Cai (1140–1195) commented, 'over half of what is needed to manage a household'.[42]

> In housing, there should be a strict demarcation between the inner and outer parts, with a door separating them. The two parts should share neither a well, a wash room, nor a privy. The men are in charge of all affairs on the outside, the women manage the inside affairs. . . . [Menservants were only supposed to enter the inner quarters to conduct repairs or in emergencies, maids were not supposed to go outside.] The doorman and old servants serve to pass messages and objects between the inner and outer quarters of the house, but they must not be allowed to enter rooms or kitchens at will.[43]

40 Tr. Ebrey (1991a: 5, 6).

41 Ebrey (1991a: 5, 11).

42 Ebrey (1984: 286).

43 Tr. Ebrey (1991a: 29). For more on the material practices of female segregation, see Bray (1997: 122–150).

The altar is the centre of this social space, which imbues all its occupants with consciousness of the proper relationships, *wulun*, and of corporate belonging. No junior member of the household can act without asking permission of the head of the family; all family members are urged to moderate their desires, curb their emotions. The family line must be continued and the members brought up as moral beings. Excess in any form is to be avoided, moderation in consumption, in sexual activity and in emotion is the key to harmony and longevity. This orthodox representation of domesticity concentrates on spatial relationships and movements, not on the quality or profusion of objects – the desire for consumption and possession is to be suppressed. Ritual and moral effects can be produced by simple means – emphasis is placed on the Confucian virtues of austerity, economy and self-control in the interests of the community. Postures are perhaps second nature but they are not natural – nobody may slouch here, nobody is ever alone or private in the modern western sense. The sentiments that should reign in this *place*, this space of decorum, are veneration, *jing*, respect for elders and superiors, *xiao*, moderation and thrift, *jie*, and kindness on the part of seniors towards those under their control, *ren*. Of course the abuse of power by seniors or high-status members was common despite philosophical exhortations to kindness and consideration.

The basic social unit within this space is the married couple, whose duties and roles complement each other harmoniously; however, although they have their own chambers, the couple do not have privacy in the modern Western sense, nor do they have any rights to private property. The man shares communal rights to the patrimonial property, held in trust for future generations. The wife is urged when she enters the house as a bride to empty her marriage chests and contribute her dowry to the family coffers.[44] Communal or corporate good, *gong*, should reign over private interests, *si*, for women as well as men.

In terms of gender, this representation of space has to be understood as affecting women of different rank or women at different stages of their lives in differing ways. Ambiguities exist at every level: women should follow men as *yin* follows *yang*, yet in certain circumstances we see women represented as the pure moral fire of regeneration at the centre of the Chinese political universe – an analogy to the "angel in the house" of Victorian England.[45] Differences of class or rank between women were spatially inscribed within domestic architecture, in the location for example of sleeping quarters. The most important spatial symbolism of female rank was that of the ancestral ceremonies, where wives participated as the indispensable ritual partners of their husbands. This ritual equality was not available to all the women in the household. Only a wife could address the ancestors, for only she was presented to them as a family member when she entered the house as a bride.

44 On the allocation of rooms within the house, see Bray (1997: 114–121); on dowries, marriage chests and women's private spaces see below.

45 For example, Ko (1994) and Mann (1997).

Concubines were barred from participation in the family cult just as they were debarred from social motherhood, and a man was not permitted to make a concubine his legal wife, even if he was widowed.[46]

The domain of the women in this space of decorum was the inner quarters, *neifang*, or women's apartments, *gui*. Although in modern Western terms spatial segregation of this kind appears to be a clear exercise of male oppression, it is unlikely that it was experienced by most late imperial women in quite such simple terms.[47] The corresponding domain for men, at least for those with any pretensions to literacy or scholarship, was the study, *shufang*.[48] This, we are told by Sima Guang, was where the sons of the household were expected to spend the day if they were not out on family business. It was in the study that men pored over the orthodox legacy of philosophical, historical and statecraft writings in the hope of passing the examinations and obtaining an official post, the most glorious (and rewarding) of careers. It was in the study that talented scholars read, reflected, and penned great works of their own, elaborating and perpetuating orthodox wisdom and useful knowledge. It was also a room where a gentleman could appropriately display his discerning masculine taste in the form of fine editions or antique writing instruments. I shall return to this form of connoisseurship below.

In Chapter 1, I outline the processes by which this decorous representation of space was gradually disseminated to produce orthodoxy through Chinese society, suggesting that spatial design and practice were consciously used by elites as a tool for social engineering, much in the way that house design was used for social reform in Victorian Britain.[49] Lineage organizations that brought peasants and gentry together in joint rituals were one vehicle for dissemination. The state too played a key role in imposing neo-Confucian representations, for instance by making Zhu Xi's wedding rituals mandatory. And as Elias would have predicted, many aspiring families expressed their claims to gentility through the construction of altar rooms and of separate quarters for their women. Western visitors to Qing China frequently remarked that even peasant houses were first and foremost ancestral temples.

If we look at the philosophies of space and time inscribed in this decorous representation of space, we find that it is an architecture that ties the family both into the kin-constructed social network of the lineage, and into the polity. In spatial terms, the centrality of the altar keeps the occupants of the house constantly aware of their bonds to living agnatic relatives, whether in the same village or prefecture, or in another province. A family may trace their lineage back many generations to an ancestor who originally came from a distant place, a home village, *guxiang*,

46 See Bray (1997: 351–358) on concubines.

47 Ko (1994); Bray (1997, 2005); Mann (1997).

48 Bray (1997: 136–139).

49 Bray (1997: 151–172).

to which they therefore feel a living connection. They are also incorporated into the social web of the state centred on the capital city: the morality of the family is fundamental to the social order of the state. Although they live in seclusion women too are tied into this political order, as producers of cloth for taxes, and as the educators of their children – an interesting sidelight on the complex significance of inner and outer, *nei* and *wai*.[50] Time too has complex manifold meanings. The family sense of time is both historical (the linearity of the lineage record, *zupu*) and cyclical (as harvest succeeds sowing and generation succeeds generation). However, in Grapard's terms the knowledge inscribed in the landscape here is more than anything historical, a constant connection of the household to founding ancestors, to the imperium and to a cultural heritage, to the precepts of Zhu Xi and Confucius.

The energetic space: selfish desires?

The popularization of neo-Confucian morality and practices was facilitated by a typically Chinese process of accommodation between "orthodoxy" and "popular custom", *li* and *su* – operating in this case through the geomancy of the house, an alternative representational space that in no way denied the spatial representations of orthodoxy and decorum, yet offered the possibility of expressing different desires. In Chapter 1, I discussed alternative, energetic readings of the relations between space and morality focused upon the altar with its ancestral tablets. We should recognize, however, that the geomantic architecture was in a sense bi-polar; although matters pertaining to the siting of the altar came first in geomantic works, in practice a geomancer was also expected to pay careful attention to the siting of the stove and its attendant Kitchen God.[51]

The inscription of gender roles and values was not especially prominent in this representation or style of being-in-the-house. As discussed in Chapter 1, geomancy affirmed that good design could *produce* neo-Confucian virtues and correct social relations, including the good behaviour of daughters and the fertility of wives. Geomantic discourse operates at the level of sex, rather than gender, in concerns with reproduction of the line and thus of female fertility. The *Carpenter's canon* contains a section on the construction of furniture in which the only item whose measurements are prescribed according to cosmological rules is the marital bed (Figure 4.1); the *Canon* also includes auspicious dates for its installation in the bedchamber.[52]

50 See Chapter 3 on the significance of cloth, and Bray (1997: 358–368) on children as ambassadors of the inner quarters.
51 Bray (1997: 107–114).
52 Ruitenbeek (1993: 226).

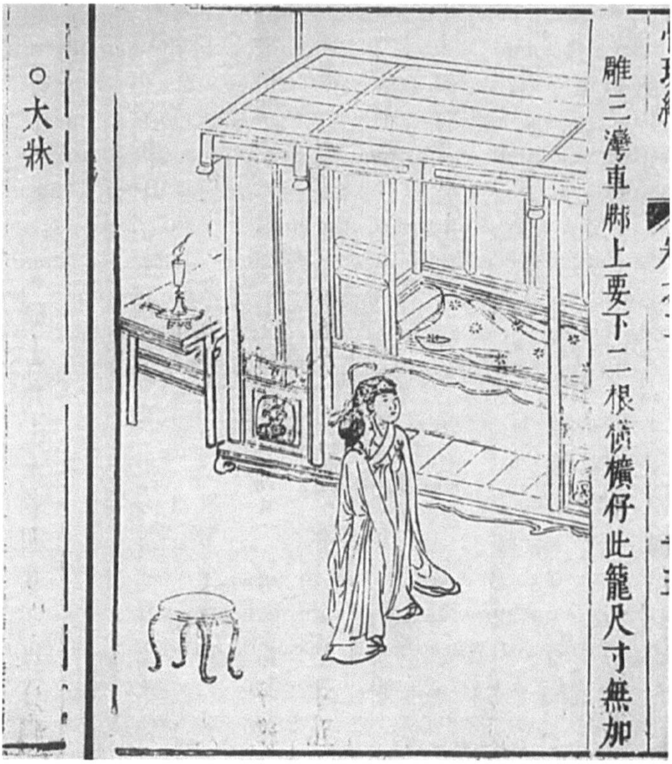

FIGURE 4.1 The marital bed; *Lu Ban jing* (Carpenter's canon), edition of c. 1600 as reproduced in Ruitenbeek (1993: 2/15b).

The textuality, and the easy legibility, of the geomantic architecture is notable: first there is the existence of the *Carpenter's canon* itself, whose widely-known maxims were built into the fabric of the house, and the spread of geomantic manuals and of almanacs with sections of geomantic advice.[53] Second, the house itself served as a complicated textual talisman to bring good fortune: matching calligraphic scrolls, *duilian*, inscribed with famous poetry, moral precepts or mantras of good fortune, were pasted up even by illiterate families, along with textual charms. Moreover, the use of punning allowed the translation of lucky characters into visual symbols, which pervaded house space as decorative motifs whose significance even the unlettered, women and children could understand.[54]

Geomancy allowed a materialist and selfish tinge to Confucian desires. It embodied a yearning for wealth and success, a readiness on the part of a family to

53 He (1990); Ruitenbeek (1993); Smith and Kwok (1993); Knapp (1998).

54 Knapp (1998).

compete against the outside world rather than working for the harmony of the community. Compared to the more Protestant and individualist trends ascribed to modernizing Europe, however, in China the importance of family solidarity was inscribed, the merits of independence, privacy or individuality were not.

Time and space in this representation were inseparable: in geomantic terms a site, *xue*, is also a moment. The house was tied to the family graves, and to a competitive local landscape in which one family's energetic benefit was often another's loss – what one might call familial rather than individual selfishness.[55] The most important form of time was familial, reproductive time, the growth and prosperity of the household within which individual bodies were born, grew and died. The rituals that accompanied these events involved eating, not the somehow insubstantial offerings of wine and tea prescribed by Zhu Xi, but meat, rice and liquor that were divided up and consumed by family and relatives according to their degree of kinship. In this story the altar depends on the stove, and indeed the Kitchen God was known to hold the destiny of the family in his keeping, as he reported on their behaviour each New Year to the celestial gods.

The most prominent desire manifested through the geomantic architecture was for good fortune embodied in wealth, health, social advancement and successful reproduction. The manipulations of environment were something one owed to the family line, yet at the same time they operated in the here-and-now. Materially, however, a house was a structure of long duration: geomantic modifications were long-term both in production and effect. Although a tree might be felled or a screen wall built to deal with immediate problems such as sickness, failure to conceive or loss of fortune, by and large geomancy was not a response to *risk* in the sense of the immediate uncertainties and dangers of a competitive and unstable society, such as Kendall proposes in her analysis of contemporary Korean shamanism.[56] Nevertheless, we may suppose that the elaborations and popularizations of this representation that occurred in the Qing, with the growth of geomantic literature and the spread of geomantic knowledge, were connected to the concomitant growth of markets, competition and entrepreneurial activity.

The aesthetic space: the amorality of beauty and ingenuity

In 1671 Li Yu (1610–1680), a well-known playwright, erotic novelist and aesthete, published *Casual expressions*, a set of essays on the elegant lifestyle. Li Yu was a sensualist with far-from-orthodox views on what mattered in life:

> For Li Yu, all pleasure is derived from an innate faculty which includes the primary instinctual pleasures or drives such as hunger or sex. The indulgence

55 Feuchtwang (1974); Bennett (1978).

56 Kendall (1996).

> of one's nature . . . should not "offend against human feeling", that is to say against a decent regard for the feelings of others; but with this one reservation, the enjoyment of beauty and sex is mandatory.[57]

Though unique in his talents, Li Yu was not alone in his views. In his rejection of puritanical Confucian values he was a typical example of what was called a "literary man", *wenren*. Literary men had always existed, but the predisposition became a way of life for many educated men during the late Ming and early Qing, when there was a dearth of opportunities for conventional careers and therefore a moral justification for the rejection of public service, and many educated men threw themselves into the arts as others might into moral philosophy or politics.

Casual expressions contains a cornucopia of musings on dilettante topics, including the composition and production of plays or opera and female beauty (sections which Hanan suggests may have been contributed by Li Yu's concubines); gardening; eating and drinking; and maintaining health (including sections on dispelling melancholy and curbing excessive sexual desire). Books 8 and 9 of the 16-book work are devoted to houses, and they represent what Hanan calls the summation of a 'lifelong passion for houses and gardens': Li Yu had designed numerous houses and gardens for patrons and friends, and for himself the famous Mustard Seed Garden in Nanjing, on a plot purchased in 1668.[58]

Ruitenbeek remarks that, along with the *Zhangwu zhi* (Record of superfluous things), a book on gardens and interior design written by the painter and collector Wen Zhenheng (1585–1645) a few years earlier, *Casual expressions* documents a development in which 'architecture was for the first time added to the list of subjects which were seen as worthy of the attention of the literary man'. Wen's book was a guide to connoisseurship that presumed a taste formed by superior education, but also the money to purchase.[59] Li Yu seldom had any money, but he came up with innumerable ingenious ways to live an aesthetically rich life for next to nothing. *Casual expressions* describes all sorts of curious or comic gadgets (adjustable eaves for poor scholars who could not afford to build a proper veranda, rocks disguised as mountains to set off a tiny garden, a bamboo pipe inserted in the wall of the study so a writer carried away by inspiration would not be interrupted by calls of nature) that Li Yu claimed to have invented as solutions to poverty or poetic rapture.[60]

Li's writings on the house are divided under five main headings, each with several sub-headings. The first section, 'Buildings', includes the following sub-sections:

57 Hanan (1988: 59).

58 Unfortunately this oeuvre vanished long ago (Hanan, 1988: vii, 29).

59 See Ruitenbeek (1993: 30) on the new attention to architecture; Clunas (1991) is a detailed study of the emergence and meanings of writings on connoisseurship in the late Ming, centred on the *Record of Superfluous Things*.

60 *XQOJ*: 8/6a–b; 18a–20b; 10a–11b.

orientation, paths, height, depth of the eaves, ceilings, brick floors, sprinkling and sweeping, and disposing of dirt. The second section, 'Windows and verandahs' starts with a sub-section on lattices and frames, followed by a fascinating essay on "borrowed scenery", *jiejing*.[61] The third section, 'Walls', has sub-sections on boundary walls, parapets, the walls of main rooms, and of the study or library, the status symbol par excellence of the period. The fourth section is devoted to calligraphic scrolls and inscriptions, the fifth to garden decor ('Mountains and stones'). Li is completely unconcerned with the ritual, magical or moral dimensions of a house. He is passionately absorbed in its aesthetics, in the fine grain of textures, in human proportions, in the relation between domestic and wild, house and landscape.

Although Li Yu in his deliberately cultivated character of the aesthete offers perhaps the most extreme expression of such representations, many of his aesthetic concerns mirror popular predispositions – for instance, it has been remarked that it is often difficult to separate geomantic from aesthetic reasoning:

> The shape of a house is of the highest importance. If the shape is unfavourable, the house will be hard to live in definitively. Whether a house is favourable or unfavourable, unlucky or lucky, can be told by the eye. As a rule, a house is favourable if it is square and straight, plain and neat, and pleasing to the eye. If it is too high and large, or too small and tumbledown, so as to be displeasing to the eye, then it is unfavourable.[62]

Vain or ambitious families might construct houses with high roofs and pillars as a reflection of their self-esteem, but Li Yu remarks sourly that while high ceilings and wide eaves are an unmistakable sign of grandeur, such splendid rooms are only comfortable in summer. In winter the owner can no doubt afford thick furs, but a poor guest will shiver in his padded jacket. On the other hand the tiny house of a poor scholar makes the most carefree visitor sigh and feel oppressed. 'Houses should be built to human scale'.[63]

Li Yu's representation of the house omits any mention of the ancestors or the altar, of female segregation, of the social separation between family and servants.

61 A common technique in garden design, which consisted of incorporating into the composition views of landscape features located in neighbouring gardens or the broader environs.

62 From a geomantic manual of 1629, *Bazhai zaofu zhoushu* (Complete book of creating good fortune in houses of all directions), tr. Ruitenbeek (1993: 38). Feuchtwang (1974: 142–143) discusses the coincidences and parallels between the auspicious and the beautiful in the selection of sites, pointing to the importance of adjectives like "beautiful", *mei*, and "graceful", *xiu*, in describing good sites. Geomantic concerns, he says, are integrated even into the most popular levels of spatial experience, 'affect[ing] the very perception of the reality of a landscape and of fortune'.

63 *XQOJ*: 8/1b.

His wife, his concubines, his servants are all complicit in his ingenious and charming inventions – not only does the servant cut the frames for his "picture window", he also joins Li Yu in laughing with delight at the result.[64] In his novel *Roupu tuan* (The carnal prayer-mat) Li Yu explicitly derides the orthodox belief that female virtue can be imposed through spatial separation and control, or male virtue produced through study of the moral classics. One apparently modern aspect of Li Yu's representation of domestic space is his concern with privacy. 'The [compound] wall separates us from others', allowing Li Yu to conduct his very un-Confucian life in peace, admitting only his friends. No ancestors frown down from the mantelpiece, and like Baoyu, the dilettante hero of *Honglou meng* (The dream of the red chamber), Li Yu and his womenfolk mingle at ease, sharing their pleasures, combining their ingenuity, adding taste to taste and distinction to distinction.

For Li Yu, the house was an opportunity to create and enjoy pleasure, to express one's talents, to turn the straw of poverty into creative gold. Li Yu's representation of the house is as an artist or inventor's studio, an attempt to shock apathetic conformity and turn poverty into an art form. A poor man can "borrow" a more fortunate neighbour's view, or create a miniature landscape on his own windowsill.

Supreme aesthetic pleasure requires originality and ingenuity; beauty alone is not enough. Here again Li Yu was the exemplar of the late Ming *wenren*, who as Clunas reminds us were instrumental in the commodification of taste: they wrote and sold books instructing people with money how to convert it into cultural capital.[65] Such *wenren* had material interests as well as social ambitions that they would have considered it poor taste to admit to. Li Yu made a good if erratic living out of selling good taste and refinement to others: his books were best-sellers, he designed houses and gardens, and he sold fine writing paper decorated to his own design. Li, who prided himself on enduring poverty, depended materially on the rich and their yearning for beauty and taste.

Although Li Yu extended his aesthetic energies to the whole of the house including the garden, the surrounding walls, and the views into the neighbours' gardens, the study or *shufang* is at its heart. The orthodox representation of the study depicted it as a place of discipline where approved books were studied and high morals absorbed, but for Li and other *wenren* it was a place that should be so designed as to put one in the frame of mind for literary and artistic creativity, for poetry, painting and music.

64 *XQOJ*: 8/20b; Bray (1997: 87).

65 Clunas (1996, 1998). Late Ming guides to good taste often advised their male readers to shun ornate or flowery styles, supposedly fit only for women or vulgar folk (Wen Zhenheng, 1996), but it seems that motifs like phoenixes and dragons were in fact favoured by male consumers even in literati families (Evarts, 1996).

In many ways Li expresses values and desires that one associates with the aesthetes of nineteenth-century France and England. Both point to the cruelty and hypocrisy inherent in orthodox ideology, and vehemently reject dominant values and goals, substituting the cult of beauty, originality and individualism – and yet, while they repudiate the accumulation of possessions, they depend on the material for a particular sense of self, which could only emerge, I think, in an age of popular thirst for the acquisition of possessions.[66]

Li's concern with the material as fulfilment and as signifier of personal worth reminds one of how the modern European greed for material extravagance was made acceptable through the concept of taste, a process that continually preoccupies French novelists such as Balzac and Flaubert. Where will the fine line be drawn between style and vulgarity? The aesthetic senses are constantly aware, any departure causes actual pain – Li Yu says the sight of untidiness is as painful as a thorn in the eye.[67] Profusion or the modern forms of bodily comfort are not pleasing in themselves. The true pleasures are those of the disciplined sensual aesthete.

Finally, the domestic space portrayed by connoisseurs such as Li Yu or Wen Zhenheng tied its occupants into a country-wide geography of local specialities, of exquisite materials and outstanding craftsman families,[68] while the appreciation of antique objects tied them into what we might call a cultural history of China, a sense of history quite distinct from that created by consciousness of lineage or loyalty to the throne.

The space of domestic comfort

Qing customs, first published in 1799, was compiled under the direction of the Nagasaki magistrate Nakagawa Tadahide, in order to provide information about China to local officers dealing with the Chinese merchants using their port. The work was also intended to appeal to a broader audience of Japanese intellectuals, hungry for knowledge of the wider world.[69] Two officials spent a year working with seven Chinese merchants to put together a picture of the manners, customs and everyday life of "ordinary people" in China. The accounts were translated by naturalized Japanese of Chinese descent, and the illustrations were made according to the merchants' directions by two Japanese woodblock artists trained by Chinese

66 Brook (1998) is one of many historians who argue that by the late sixteenth century China was well on the way to becoming a consumer society.

67 *XQOJ*: 8/10a.

68 Clunas (1991); Brook (1998).

69 Under the Tokugawa shogunate Japan was closed to foreigners and Japanese were forbidden to travel abroad. The port of Nagasaki was the only place where foreign traders were allowed to set foot on Japanese soil.

masters.[70] The seven merchants, like the translators who worked on their accounts, came from four cities in the Yangzi Delta (three from Suzhou, one from Huzhou, two from Hangzhou and one from Jiaxing), and they were also familiar with the port cities of Fujian. But as they told Nakagawa: 'China is a very large country whose customs differ greatly from province to province. We come from the Southeast, and can only describe the customs and scenery of that region'.[71]

The book is in 13 sections: yearly activities (including festivals); the house; clothes; food and drink; education; childbirth; the rituals of capping and pinning (which marked boys' and girls' transition to adult and thus marriageable status); marriage rituals; receiving guests; how to travel in China; funeral rituals; Confucian sacrifices and rituals; Buddhist and Daoist monks. The section on the house is by far the longest, 75 double pages in length, and copiously illustrated.

Here we have the only systematic descriptive account of domestic space from late imperial China of which I am aware. The section on the house is organized as a systematic movement through the house, entering at the front gate and proceeding to the rear courtyard, with systematic enumeration and illustration of building forms and spaces, their uses and the material objects that typically would be found in any room. Some intimate objects, such as sanitary napkins and chamber pots, are not, as far as I know, illustrated and described in this way anywhere else in the Chinese record. It is almost as if a late nineteenth-century German ethnologist had been systematically at work here – and no doubt we are seeing the influence of Japanese forms of depiction of material life in their own society translating into curiosity about the details of life in China.[72] But I think we are also seeing a distinctive Chinese vision of the house, enhanced perhaps by nostalgia. *Qing customs* depicts a largely secular, materialistic world – a peaceful and not especially rigid life of pleasant relations between family members of both sexes, their circles of relatives, colleagues and friends.

Although both Nakagawa and the merchants were anxious to show respect towards the venerable and sagely traditions of Chinese civilization, it is striking that Confucian ceremonies are left to the penultimate section on sacrifices and rituals, where they jostle with notes and illustrations on the cults of the City God, Tianhou and Guandi, as well as a charming illustration of a temple opera. The last chapter covers Chinese Buddhism, with a final page and picture of a Daoist priest as a kind of afterthought – is this simply a response to Japanese ignorance of Daoism, or did

70 Some of the illustrations were copied directly from Chinese sources such as the *Sancai tuhui* (Illustrated encyclopedia of Heaven, Earth and Man) of 1607 (see Chapter 8), but all the double-page illustrations were completely new.

71 *SZKB*: 13.

72 Sets of scenes of daily life and crafts in Japan were common by the latter half of the Tokugawa period. They also appear in South China quite early in the nineteenth century, but these were intended principally, it seems, for Western purchasers.

it play little part in the life experience of these very urban Jiangnan merchants and their families? We can see that although they were comfortably familiar with ritual and propriety, still the altar was not as central in the minds of these respectable merchants as orthodox Confucians would have wished. We are hardly surprised to find that Li Yu omits any mention of family altars in his essays on the house. But consider the description of the main hall given in the *Qing customs* in the early chapter on the house:

> As soon as one is past the inner gate, one reaches the reception room for visitors. It varies in size and has two or four-panelled folding doors, a tiled floor and pillars with stone bases, round, square or hexagonal. *At the centre of the back wall there is an alcove; in front stands a long-legged table. A framed painting or piece of calligraphy hangs above the table, and there are paired scrolls hung on either side. On the table are incense burners and flower vases.* Chairs of rosewood or rattan are placed on either side of the table, and guests are invited to sit here and chat. Doors open out of the back wall to allow food to be brought in from the kitchen.[73]

It is clear that the highlighted passage refers to the shrine, and the accompanying illustration (Figure 4.2) confirms it, yet no mention is made here of the ritual significance of this "alcove". Only in the later chapter on sacrifices is the shrine shown as a place of worship.[74]

In that chapter *Qing customs* enumerates the yearly round of domestic offerings to the ancestors. The dates correspond to those of the *Family rituals*, but the spirit seems different from the rather impersonal respect, *jing*, of neo-Confucian orthodoxy. The work conveys a sense of a more intimate relationship between family and ancestors. The ancestors receive personal attentions on their birthdays, when their portrait was hung over the altar and received special offerings.[75] There is also a loving attention to the details of the offerings. A routine offering might include lychees, candied melon and sweet cakes. The drawing of the main annual sacrifice (see Figure 1.2) is so detailed that it would not be difficult to reproduce the feast, from the three roast animals (pig, chicken and sheep) to the dishes of rice and bamboo sprouts.

The representations of gender are also much less patriarchal than those of strict orthodoxy, though they are strictly respectable. Spatial segregation appears a little lax by strict orthodox standards, though there are clearly standards of propriety at

73 *SZKB*: 75–76, emphasis added.

74 *SZKB*: 498.

75 Orthodox Confucians objected to the use of portraits for such rituals on the grounds that they could never be sufficiently accurate.

FIGURE 4.2 'Front view of the reception room', an informal view of the family altar; *Shinzoku kibun*.

work determining how far a particular guest or relative penetrates into the building and whether or not the family women will take part in entertaining. Although we are told that in poor homes female modesty sometimes had to be preserved with a simple curtain, the merchant houses had substantial "inner quarters" in the rear courtyard, including a separate block, two stories high, containing the daughters' bedrooms and the "dew platform", *lutai*, all tastefully furnished.[76]

The dew platform (Figure 4.3) was on the upper storey, a private room where male visitors did not go. It was mostly occupied by the women, but the men would also spend leisure time there when they could. Designed for the semi-tropical climate of Jiangnan, the room consisted of a large veranda protected from the sun by an awning that caught every passing breeze, wafting in fragrance from the plants set

76 *SZKB*: 114–115, 132–133.

FIGURE 4.3 The "dew platform", *Shinzoku kibun*. A Buddhist altar is on the left, with calligraphic scrolls on either side. The small table at the back of the room is a writing desk, with books, an ink slab, water dropper, brushes and brush stand: the scholar's "four friends".

in pots along the balustrade. A Buddhist altar stands against one wall – an alternative spiritual focus to the ancestral tablets in the main hall.[77] Fathers tell us proudly how they take breakfast or spend their leisure time with their wives and daughters on the dew platform, especially enjoying festivals such as the autumn moon viewing. For Sima Guang or Zhu Xi, such companionability would represent a promiscuous mixing of the sexes.

There are two more very interesting transgressions of orthodox gender doctrine, or at least deviations from strict orthodoxy, in *Qing customs*. The first concerns the disposition of female property. In the picture of the lady's chamber we are shown the alcove containing the dowry chests (Figure 4.4); the text tells us that wives kept these belongings locked up – that is to say they had not been obliged to give them over to the family, as required by strict Confucian orthodoxy. No mention is made of any potential conflict: it seems the Jiangnan merchants took the rights of women to control their own property for granted. Even maids, according to *Qing customs*, had a key for their private rooms.[78]

77 On Buddhism as a female cult, see Bray (1997: 133–136).

78 *SZKB*: 115–116. See Bray (1997: 139, 171) on contending claims to dowry.

FIGURE 4.4 Lady's chamber, *Shinzoku kibun*. The alcoved bed forms a chamber within a chamber. On the dressing table by the window are a round mirror, a jewellery or cosmetic case with several drawers, a toothbrush, a band for tying up the hair, a teapot and a cup. A washstand is on the right, and there is a clothes rack by the entrance to the bed. A locked alcove contains the woman's chests of clothes and valuables.

The second concerns the study of texts and literary production. At the back of the *lutai* veranda we see an elegant writing desk, furnished with all the necessary scholarly appurtenances from inkstone and pen stand to finely bound books. Clearly the young ladies of this merchant household had gone beyond functional literacy, in what might well be regarded by an orthodox traditionalist as a double infraction of class as well as gender roles. The debate over how much education women should receive involved many scholars and women of refinement through the late Ming and early Qing.[79] What is remarkable here, I think, is that these late eighteenth-century merchants do not even bother to comment on possible moral or social objections to educating their daughters, though they do tell us that they have to hire in tutors for them because, of course, girls cannot go out for lessons like boys.[80] Furthermore, they take it for granted that merchant families share the spatial and cultural appurtenances of the gentry, and show mild astonishment only when rustics lay claim to similar cultural capital: 'Wealthy farmers' houses include

79 There is now an extensive literature on this issue; see, for example, Ko (1993), Mann (1997) and Fong (2008).

80 *SZKB*: 155.

a reception hall, a study and inner quarters *just like merchants' houses*',[81] they tell the Nagasaki officials. This way of putting it would certainly have astonished the gentry elite of the early Ming, but already by the latter half of the Ming the boundaries between merchants and gentry-scholar status were becoming very permeable, and even policemen in some of the rich lower Yangzi towns were adding what they called "studies" to their houses, to the disgust of their educated superiors.[82]

Of my four sources, *Qing customs* is perhaps the representation of being-in-the-house that comes closest to the modern European ideals of bourgeois domesticity that translated into new norms of middle-class dwelling in the course of the nineteenth century. Is it merely coincidence that the tale is told by merchants? What are the effects of the Japanese input?

The pages of *Qing customs* exude material and social comfort, describing well-to-do households whose members take pleasure in each others' company and in their daily occupations, and who enhance their enjoyment of their home by surrounding themselves with material objects. In its dealings with material possessions and its representations of a culture of consumption, the spirit of *Qing customs* is descriptive, not prescriptive. We are shown a wide variety of everyday objects (pedestal lamps, reading lamps, wall sconces, and little hooded lamps for desks; stools, upright chairs, armchairs, folding chairs, couches and benches, just to mention a few) with descriptions of what materials they may be made of, where they are likely to be placed, how they are decorated.[83] But there is no suggestion, as in the late Ming works of connoisseurship, that people are judging each other by minute refined criteria of style. We are not told that the best-quality rattan stools come from X prefecture and are made by Y family, while stools from Z are undesirable. Nor do we find here – as in Li Yu – that sensual enjoyment is a philosophy and a form of discipline. Rather, a cheerful profusion reigns.

The representational spaces depicted in *Qing customs* are sociable: the inhabitants receive visits from neighbours and family, affines as well as agnates. Clients and colleagues call, the boys go out to school, the tutor comes in for the girls, the midwife and the wet nurse remain in touch. Time is marked out in years by the cycle of festivals, by the birthdays of the ancestors and feasts laid out in front of their shrines. Like Zhu Xi, this work emphasizes rites of passage, and yet it differs in significant ways: the ceremonies of capping and pinning and of marriage are preceded by a chapter on childbirth. Meanwhile the section on funerals and mourning rites – the core of neo-Confucian ritual concern – comes separately,

81 *SZKB*: 181, emphasis added.

82 Bray (1997: 138); Brook (1998).

83 Although many images are close copies of originals from the *Lu Ban jing* or *Sancai tuhui*, many others are new, while many objects that featured in the earlier texts are not included.

almost at the end. The orthodox ties to history through state or lineage, the spatial ties to imperial capital or to ancestral provinces are scarcely apparent; instead these families are active city dwellers, they visit temples, frequent operas, go on pilgrimages, while the merchant heads of families travel far and wide and command a portfolio of knowledge and etiquette in consequence.

Domestic spaces and paths towards modernity

In this chapter I have suggested the elastic capacity of late imperial domestic architecture to accommodate diverse structures of feeling within an overarching framework of orthodoxy. But beyond elasticity, can we discern any significant historical trends? Historians of Europe attribute considerable importance to the role played by spaces and their representations in the formation of a modern society and of modern identities, sensibilities and desires. It is now quite commonplace to treat late imperial China as a society embarked on its own path(s) towards modernity. Whatever the similarities with social, economic or political trends in early-modern Europe, clearly the changes in China did not produce what historians of Europe identify as "modern individuals" – if indeed such beings exist. However, I think that we can point to some important historical shifts in sensibilities and ideologies of domestic space, some of which do offer intriguing analogies to those that occurred in Europe.

The four vignettes presented here were chosen because they provide four vividly contrasting representations of space, all housed within the same basic framework of pillars and beams, wings and courtyards. In date of composition the four texts span almost a millennium. I am not suggesting that the order of these four different representations implies a historical sequence, a story of development or of definitive shifts of ideology or sensibility. Nevertheless, it is significant, I think, that a full exposition of the space of decorum is the first to appear, and it seems reasonable to propose certain historical trends in the balance between these different representations.

The orthodox neo-Confucian "story", as inscribed in the neo-Confucian writings of the Song, represented the house as primarily a space of ritual and etiquette, a stage for the performance of the correct family and reproductive relations that underpinned the social order, a domain of moderation, responsibility and unselfish virtue that tied its occupants into a web of patrilineal kinship, into the wider polity, and into a moral tradition. This story of the house served as a means of disseminating orthodox values, not least because (as suggested in Chapter 1) it provided the symbolic tools for transforming material diversity or inadequacy into ritual conformity, so that all houses, whatever their design, could serve as the proper stage for orthodox performance. This particular "story" was initially known to a rather small proportion of Chinese society, namely the scholarly elite. As this group extended its influence over local society they circulated the story, in particular through ritual leadership and through the establishment of lineage networks that included kin-groups spanning the economic classes. Increasing numbers of ordinary families became acquainted with this architecture of decorum and claimed it for themselves. Shared spatial practices were thus a powerful form of acculturation that contributed importantly to diffusing orthodox values throughout society.

The orthodox architecture of decorum, an embodiment of neo-Confucian principles, played an analogous role in late imperial society to what Lefebvre calls the "scientific" representation of space in modern Europe, gradually spreading through the social ranks until it became common sense or second nature, even in humble families. But the dissemination of this form of spatial consciousness coincided with other social trends, notably the progressive shift from a subsistence to a commercial economy. This began in the Song, with growth in the size, number and wealth of cities and attempts to monetize the tax system. The trend was not linear: wars and climatic disasters brought intermittent collapses, while the first rulers of the Ming and Qing tried valiantly to restore what they saw as a more natural and virtuous social order of self-sufficient villages and urban frugality. Nevertheless, the long-term trend, spearheaded always by the Jiangnan provinces, was towards a money economy in which inter-regional trade flows and the demand for handicrafts and manufactures increased, farmers traded what they grew and bought food for their own consumption, urban populations and the number of well-to-do households increased, and the boundaries between merchant and landed gentry became increasingly permeable.

The market always carries risks, and as more and more families came to participate in the market they faced competition, the need to make decisions, and the possibility of ruin. In such a context it is not surprising that the geomantic, energetic architecture of the house came to prominence, with excerpts and drawings from the geomantic and building-magic sections of the *Carpenter's canon* widely featured in popular encyclopaedias and almanacs. This representation of space, emphasizing the fulfilment of such desires as wealth and success in business, accommodated an ethos of competition and selfishness that the architecture of decorum played down in favour of harmony and frugality.

Patriarchs throughout the late imperial period continued to pen family instructions in which they deplored waste, lavish expenditure, or economic ventures that might put the patrimonial estate at risk. But reality was somewhat different: throughout society, levels of consumption rose and markets for what had once been luxuries expanded. Taste itself became a commodity as a literature of connoisseurship developed, and the architecture of aesthetics I have illustrated with Li Yu's work represents the spatial possibilities both for refined sensual pleasure and for the expression of cultivated individuality. At a less exalted level of refinement and investment than that represented by the collection of rare antiques or perfect lacquerware, as *Qing customs* suggests, growing numbers of people partitioned off part of their house as a study, bought books, brushes and carved inkstones, and arranged pots of orchids and carved stones in front of the study window. At the level of comfort and convenience, the profusion of furniture types and household objects illustrated in *Qing customs* is heralded by entries on household furnishings and equipment in encylopaedias throughout the Ming and Qing. More and more well-off families grew accustomed to taking comfort and material profusion for granted.

The analogies of spatial consciousness and forms of identity between late imperial China and modernizing Europe that I have pointed to in this chapter seem (not surprisingly) to spring principally from the fact that both regions experienced long-term economic expansion and commercialization, and as a result both saw the development of an increasingly prominent role for the merchant class, with growing possibilities of assimilation into the elite. But is there not one striking and significant contrast, namely in the place and roles assigned to women? How does the confinement of Chinese women within the inner quarters affect my arguments?

I have already suggested that the overlapping architectures sketched out here accommodated a wide range of interpretations of gender roles and relations. Even within the overarching patriarchal ideology of orthodoxy, two starkly contrasting interpretations of spatial segregation were possible. Either women were morally inferior by nature to men and must be sequestered, controlled and protected from their natural propensity to vice; or women were morally equal to men, their indispensable ritual and social partners, and because spatial seclusion protected them from the pressures and temptations that men inevitably encountered in society, they were able to cultivate untainted and superior virtues, which they inculcated into their children to the benefit of all society. *Casual expressions* and *Qing customs* evoke different kinds of relaxed and pleasurable companionship between men and women within the family, and both take female literacy for granted, for ladies if not for maids. *Qing customs* also takes it for granted that women had rights to private spaces, and that wives retained personal control over their dowry wealth, as indeed they had in some families for centuries past. Some wives in late imperial China had not a copper cash to call their own, others took control of family finances, including investments and management of family farms and businesses.[84]

Yet Chinese women were spatially confined – and by the late nineteenth century this was viewed by both Western critics and Chinese reformers as a key cause of national weakness, an obstacle alike to intellectual and economic modernization.[85]

84 Ebrey (1993); Bray (1997: 358–368) on the wifely role, and Bray (2005) on the paradox whereby wives had claims to private spaces and property in imperial China, where their husbands' individual rights were subordinated to the patriline.

85 See Ko (1993) on "May 4th stereotypes".

How, in fact, did China compare to European societies on this score? Certainly there is truth to the contention that in Europe new architectural forms and norms promoted and naturalized concepts such as individuality and privacy; they also helped propagate norms of healthy family life, efficiency in housework and strict gender divisions of labour. Yet, as in the case of China, multiple representations and interpretations of domestic space overlapped or contended, contradictions and paradoxes abounded. In Britain, not until 1870 did the Married Women's Property Act allow wives to retain control of their dowries and their wages. And not only under Victoria's reign, but well into the 1960s, female seclusion in the persona of the housewife was consecrated as the necessary foundation for a productive economy and for the confident manliness that kept the Empire going.

In other words, the differences in emerging spatial ideals in early-modern Europe and China are, I think, significant – but not because they represent an obdurate tradition that impeded China's progress towards modernity. On the contrary, the contradictions and divergences in the readings of Chinese domestic space and its occupants discussed here offer a useful prism through which to investigate certain important particularities of the complex historical processes of economic, political and cultural change that marked China's trajectory through the period we might best think of as a worldwide early-modern era. Above all, I think that this analysis of how representations of domestic space accommodated the tensions generated by long-term social transformation demonstrates the importance of incorporating material culture into historical analysis.

5

TALES OF FERTILITY

Reproductive narratives in late imperial medical cases

In *The Woman Beneath the Skin*, widely considered a conceptual milestone in the history of medicine, Barbara Duden complained that the techniques of historical demography 'silence the body':

> To social historians of my generation historical demography has become a primary source of statements about body-mediated phenomena [in which] the characteristics of the body within a statistical population are – perceived as *probable attributes of an object*: as rates of birth, morbidity, reproduction and mortality.[1]

In recent years demographic historians of China have been striving to recapture what they call the *demographic mentalities* of the people whose individual births and marriages, pregnancies and deaths coalesced into the grand ebbs and flows of population change, which they seek to trace and to explain. 'Eventually we will understand not just how frequently individuals marry and give birth, but why.'[2] Is this optimism justified?

The new demographic history of China is part of a broader revisionist project. Earlier Malthusian interpretations held that in contrast to early-modern Europe, where families operated deliberate *preventive checks* (such as late marriage) to reduce the number of children women bore, thus increasing the resources available to individual offspring in an increasingly competitive society, in imperial China

1 Duden (1991: 192, emphasis added).
2 Lavely *et al.* (1990: 818).

population growth was limited only by the *positive check* of death. There was a cultural explanation, a *rationality*, underlying this statistical pattern, but it was distinctly non-modern: namely, families were driven by the imperatives of patrilineal descent to have as many sons as possible.[3] But over the last decade a new image of late imperial China has been fashioned. The main thrust of recent comparative, economic, intellectual and cultural history has been to argue that late Ming and Qing society (c. 1550–1800) was working its way, just as Europe was, towards modernity. Late imperial China was modern not in a European but in a distinctively Chinese mode, and it contributed in ways hitherto insufficiently scrutinized to the shaping of the emerging modern world, its international relations, statecraft practices, commercial institutions and patterns of consumption.[4] Not surprisingly, demographic historians have been anxious to demonstrate that late imperial population history fits the same mould.

The data-sets upon which the demographic historians of China base their models of reproductive rationality are composed chiefly of documents such as lineage records, *zupu*, and censuses.[5] Although critics still contest their use of the admittedly difficult and partial sources, the new demographers use statistical analyses of these data-sets to assert confidently that marital fertility rates in late imperial China were significantly lower than those in contemporary Europe, while births were evenly spaced across a woman's fertile period, indicating that families took deliberate measures to space their children.[6] Referring to studies by social or cultural historians, including myself, the demographic historians note as part of the context in which individuals made decisions a range of social and biological techniques for controlling the number of offspring, including infanticide; contraception or abortion; adopting in or out; and perhaps low rates of coital frequency following prevailing medical wisdom.[7] They point to the late imperial expansion of gynaecology and paediatrics as an indication that families were prepared to invest in the health and future of individual children, rather than trying for as many male offspring as possible in the hope that some would survive. They link such imputed attitudes to a broader context of social and economic competition and higher levels of affluence – in other words they see here a form of "modern" rationality framing reproductive decisions.[8]

Yet this approach to *demographic mentalities* silences not only the bodies, but also the persons most intimately involved. First, the genealogies that are among the principal sources for Chinese demographic history record patrilineal descent. Since

3 This view is still current among leading demographers in China itself (see Lee *et al.*, 2002). Arthur Wolf is probably the most passionate proponent of this view in the West today (e.g. Wolf, 2001).

4 Frank (1998); Wong (1997); Pomeranz (2000); Hamilton and Chang (2003); Elman (2005); Brook (1998); Clunas (1999).

5 Campbell *et al.* (2002); Lee *et al.* (2002).

6 Lavely and Wong (1998); Lee and Wang (1999); Zhao (2002).

7 Drawing, for instance, on Furth (1987, 1999, 2007b), Bray (1997) and Hsiung (2005, 2007).

8 Campbell *et al.* (2002: 742); Li Bozhong (1994, 2000).

Chinese women were incorporated into their husband's lineage on marriage, daughters were not usually included in these documents; wives were sometimes recorded, sometimes not; concubines were omitted. Even sons were often not registered until they had survived the first years of infancy. What womanly experiences of barrenness and miscarriage, of bearing nameless daughters or sons who perished in infancy, lie behind the bare facts of names and years noted in a genealogy? In other words, how did a woman's *fecundity*, the number of pregnancies she experienced, translate into *fertility*, the number of children who were born, survived and made it into the records? Demographic historians are able to test statistically whether certain practices or clusters of practices (such as infanticide and/or abortion) are consistent with their general model.[9] This means, however, that they can account only for actions that were efficacious in modern terms. What they cannot test for is intentions, beliefs and partial efficacy – fundamental components of the "why" and the "how" of reproductive behaviour.[10]

As an alternative source of insights into demographic mentalities, or what I prefer to call *reproductive cultures*,[11] here I offer a brief exploration of medical cases describing the treatment of reproductive problems. I agree with scholars such as James Lee that the information that such qualitative sources offer resists easy incorporation into demographic models, being highly circumstantial and concerned principally with anomalies rather than routines. At the opposite end of the analytical spectrum from abstract models, these medical cases focus on the experiences of individual women, recapturing the bodily experiences, the aches and palpitations, the yearning, elation and fears surrounding fertility and childbirth. Sometimes they span the few hours of an acute crisis, sometimes they chart constitutional changes over many years. As narratives with deliberately selected beginnings and ends, they offer insights into how female health and fertility were construed, and what was considered the desirable outcome of a reproductive disorder. Furthermore, they incorporate the supernatural explanations and the social expectations that were as important as economic rationalities in shaping reproductive choices and reproductive cultures.

Medical cases in late imperial China

The late imperial period saw the emergence and spread of a prosperous, literate urban culture that supported a burgeoning of medical practice and medical

9 For example, Campbell *et al.* (2002).

10 Etkin (1988); Bray (1997: 290–296).

11 The term *mentality*, developed by *Annales* historians in the Durkheimian tradition, denotes common ways of thinking or casts of mind that prevail throughout a society. Anthropologists and social or cultural historians today treat *culture* as a constantly shifting arena of contested definitions and conflicts over power.

publishing. Literati had always had some familiarity with medical theories as applications of general cosmological principles. Furthermore it was the duty of a good Confucian to prepare medicines for ailing parents, and many famous literati physicians through the ages first learned their skills in order to treat sick family members. By the late Ming, many ordinary families could afford to consult doctors. The competition between physicians, both for clients and for clinical authority, was intense. Some identified themselves as hereditary physicians, *shiyi*, claiming generations of skilful doctors as forebears; others identified themselves as *ruyi*, literati physicians, their virtuosity rooted in superior learning. Religious and magical beliefs produced variations on or alternatives to medical images and interpretations of the body, and orthodox physicians competed not only with each other but also with folk healers, midwives, diviners, bonesetters, priests and herbalists.[12] Most people who consulted doctors had at least a smattering of medical theory, gleaned from the medical sections of almanacs or household encyclopedias, or from cheap primers on medical topics written in accessible language. Many felt competent to question a doctor's decisions and call in a rival. There was a huge market for every kind of medical work, and new editions were sometimes pirated within days.[13]

The Ming publishing industry contributed significantly to the professionalization of medicine: a dense web of technical debate and argument over claims to truth was made open to public scrutiny.[14] Argument through cases had played a key role in the construction of knowledge in both law and medicine from early times, in China as in Europe. But individually authored collections of medical cases, *yian*, were a Ming innovation, closely connected to the rise of the *ruyi* or literati physicians.[15] Collections of medical case histories, with their tales of subtle diagnoses and inspired cures, fascinated fellow literati and the wider public.[16] They sold well as literature,[17] while helping physicians to establish convincing claims for their preferred theories and therapies: they 'turned attention away from inherited canon and ancient masters, locating medical authority in the individual'.[18]

Medical cases speak primarily through the authoritative voice of the doctor. But they are also narratives of sickness and treatment that allow the sick persons and their bodies to speak, even if indirectly. Some physicians recorded cases in very concise terms, occasionally reducing them simply to a set of symptoms and prescriptions. More commonly, cases told a human as well as a medical story. Most

12 Chao (2009: Chapter 1); Cullen (1993).

13 Chao (2009: 28); Chia (2002).

14 Chao (2009); Furth (1999: 224–265, 2007a, 2007b).

15 Furth *et al.* (2007); Cullen (2001); Grant (2003).

16 On the similarities in structure and allure between medical cases and detective stories, another genre that became wildly popular in the late Ming, see Zeitlin (2007).

17 Furth (1999).

18 Furth (1999: 226).

late imperial cases highlight the experiences of individuals embedded in their social circle: physicians did not have surgeries but visited clients in their homes (often in between visits from rivals); they consulted with husbands, mothers, maids and family friends; they made repeated visits; they recorded not only the details of bodily sufferings but also the emotional triggers to which they attributed physical dysfunction, and their negotiations with family members as the treatment progressed. Many cases describe a patient who has undergone previous diagnosis and treatment by other physicians or healers – naturally the earlier treatment was mistaken and unsuccessful, while the virtuosity or subtlety of the physician who recorded the case brings about a cure. In some cases the patient or their family ignore the advice of the recorder of the case; the results tend to be fatal.

A case is a professional construction, a selective narrative that retrospectively imposes order upon confusion, that disentangles essential from coincidental, primary from secondary. Where do the roots of the sickness lie? What does the patient think is wrong, and does it coincide with the doctor's understanding? What are the stages of treatment, and at what point has the problem been resolved? The choice of where a case begins and ends is not arbitrary, and may well require us to reflect upon the distinction that anthropologists make between *curing* and *healing*. Curing is defined from the practitioner's perspective, as the elimination of a pathogen or the repair of a lesion; healing is defined from the perspective of the patient, as the resolution of the disorder from which they perceive themselves to suffer. Thus if someone has a raging fever that they attribute to possession by a demon, a biomedical doctor would expect to *cure* them by administering antibiotics to eliminate the infecting virus, while the sufferer would consider this step to be the suppression of a mere symptom, believing that the fundamental problem of possession could only be *healed* by an exorcism.[19] Medical theories in late imperial China varied considerably, and none had the degree of authority claimed by biomedical science today; cases sometimes show prolonged negotiations between physician and client about what exactly was wrong. Most medical cases were written up for public as well as professional readerships, and were thus crafted to demonstrate that the doctor could bring about happy endings for his patients – as long as they were wise enough to follow his advice. Those who foolishly ignored it suffered the unanswerable alternative to a happy ending: death.

Womanly health

One medical specialism that expanded rapidly during the late imperial period was "wives' medicine", *fuke* (or "women's medicine", *nüke*). The notion of a specialized female medicine first emerged in the writings of some Song dynasty physicians

19 Nichter (1992).

such as Chen Ziming (1190–1270), in response to the idea that the different endowments of *yin* and *yang* characteristic of male and female constitutions produced a sexual difference in the responses and processes which patterned a disorder, and required corresponding differences in prescription. The earliest works of *fuke* therefore included a full range of general disorders as they affected women, as well as special sections on what from a biomedical perspective we would think of as gynaecology or reproductive medicine.[20] By the later Ming, however, the category of *fuke* was devoted almost exclusively to the disorders of menstruation, conception, pregnancy, birth and its aftermath; in other words, it coincides rather closely to the modern biomedical category of gynaecology.[21] Sections on *fuke* feature prominently in late imperial collections of cases, and here I present cases from some well-known Ming and Qing specialists. By charting anomalous experiences that warranted medical attention, *fuke* cases offer insights into what kinds of reproductive experiences women or their families felt were normal, desirable or tolerable – at least among the comfortably-off or rich urban families who formed the bulk of the *fuke* physicians' clientele.

Most of the cases I present here come from the chapter on *fuke* in a critical compendium of cases published in the Qianlong reign (1736–1796) by Xu Dachun (1693–1771). Xu was an archetypal *ruyi* or literati doctor. He began to study medicine because many of his relatives fell sick, and although he gained renown as a practitioner, he was still more famous during his lifetime and after his death as a learned scholar and prolific writer. The chapter on *fuke*, the thirty-second in the work, is long (39 double pages of small print). It includes numerous cases from Xu's own practice, and many from famous Ming physicians including Wang Ji (1463–1539)[22] and Xue Ji (1487–1559).[23] I also cite a few examples from Ye Gui (1667–1746), whose cases were collected and published posthumously by his disciples in 1776. Ye and Xu were both natives of cities in the prosperous Lower Yangzi region. Unlike Xu, however, Ye considered himself a hereditary physician; he started learning from his father and apprenticed himself to 17 masters before setting up his own practice. Wang Ji and Xue Ji also learned their craft from their fathers.[24] The roughly 300 cases in Xu Dachun's *fuke* section vary in length from a couple of columns to almost a page. Ye Gui's cases are unusually long and detailed. In both collections the patient is usually not named, although age and marital status are almost invariably included; not surprisingly, the great majority concern married women of childbearing age, though pubescent girls and older women also feature.

20 Furth (1999: 59–93).

21 Wu (2010) develops the parallel further, emphasizing the paternalism inherent in both medical specialisms.

22 See the study by Grant (2003).

23 The renowned author of numerous works, including theoretical treatises on *fuke*.

24 Ou (1988: 354, 361); Li Jingwei (1988: 97, 281, 668).

One difficulty with assessing the cases from the perspective of reproductive cultures is that sometimes the term son, *nan*, or daughter, *nü*, is used for a baby, but more often the physician simply uses the term child, *zi*, a term that in principle does not denote sex yet in the context of patriliny often implies a male child or heir.

The *fuke* cases offer some interesting insights into how female fertility was construed by non-specialists. It was standard medical wisdom, dating back to the ancient *Yellow Emperor's Classic of Internal Medicine*, that a girl's fertile energies come through at twice seven years, *sui*, and a boy's at twice eight; at that point both are able to procreate. Women's fertility is exhausted at seven times seven *sui*, men's at eight times eight.[25] The Song physician Chen Ziming had advised that a woman should not bear children before twenty *sui* for fear of damaging her internal organs, while a man should wait till thirty so that his *yang* energies were firm; Chen's arguments were elaborated by Ming medical specialists and were much in vogue among physicians in the eighteenth century.[26] The *fuke* cases appear to confirm that in late imperial times nice families did not set girls to child-bearing as soon as they were physically ready. Out of all Xu Dachun's cases involving pregnancy, if age is mentioned at all we find that the woman is at least twenty *sui*; only in a single case (included in the section on menarche, *tiangui*) is it implied that the pregnant woman is a very young girl – indeed probably unmarried. A woman brings her daughter (age not given) to Xu Dachun who examines her and finds a pulse that indicates pregnancy. He suggests that the young lady has had no period for over a month, and is told he is correct. He provides intensive care throughout the duration of the pregnancy first to encourage the healthy formation of the fetus, later to avoid miscarriage and finally to help the young lady give birth easily. 'And indeed she gave birth to a boy, who was bouncy, bright and easy to raise'.[27]

On the onset of menarche, it is interesting to note that physicians were likely to qualify a girl who started menstruating at fourteen *sui* (the classical twice seven) as reaching menarche unusually early. Other cases indicate that most mothers did not start worrying unduly about the absence of menses until a girl was about seventeen *sui*,[28] and that scrupulous physicians tried to calm over-anxious parents:

> A woman brought in one of her daughters, fifteen *sui*, to be examined. She said that when the girl was fourteen her menses started to flow on their own, but now on the contrary they had stopped. The mother told me this with

25 Bray (1997: 318). A child entered its first *sui* at birth and its second at the passage of the lunar New Year; thus a child of four *sui* would be about two years old in Western reckoning. The multiples of seven years for women and eight for men clearly have a numerological touch; nevertheless 12 Western-style for menarche and 47 for menopause seem reasonable approximations.

26 Furth (1994); Wu (2010).

27 *YLLS*: 32/12a–b.

28 For example, *LZZN*: 6/14a; *YLLS*: 32/3a.

> fear. I said: 'Madam, this is your daughter and your menses must have begun at fourteen too. The reason that they have stopped is that they have skipped a year and you find this strange, but later they will start again. However I still fear that as to True *qi* [the body's protective and growth force] her natural endowment is both irregular and weak. If you want to treat it with drugs, that would be appropriate.' I thus secured the innate True *qi* to make Water rise and Fire descend, so that the Five Viscera were in harmony with each other and the pulses were in communication.[29]

This case is typical in that the negotiation takes place between the person socially responsible for the patient and the physician, rather than directly between physician and patient. (A mother speaks for her unmarried daughter, a husband will usually negotiate with the physician on behalf of his wife.) At the level of diagnostic virtuosity, the case illustrates that sometimes the interruption of a natural process may in itself be a natural event. The physician tries to allay the mother's fears (well supported by classical medical theory) that a disruption of the menses signals the onset of some profound disorder[30] by appeal to specific personal circumstances: the mother herself probably started menstruation early, but here she is alive and well and with a pubescent daughter. Probably nature can be left to take its course. However, the mother obviously wishes to take action, and perhaps there is some slight weakness in the girl's vital force. Prescribing a tonic to replenish *qi* and Blood would certainly do no harm, ensuring the balanced physiological flows that signalled female health and fertility. It was also a means to retain mother and daughter as satisfied clients,[31] whereas refusing to prescribe in response to what they perceived as a life-threatening problem would probably have sent them to another more obliging physician, as in the case treated by Xue Ji, which Xu cites immediately previously:

> A virgin, seventeen *sui*, ill for a long time without cure. Her reproductive powers had not come through [i.e. she had not yet begun to menstruate]. She developed a fever and a cough and ate and drank little. They considered using emmenagogues, but Xue [Ji] said: 'This is because her natural endowment is not sufficient and the *yin* Blood hasn't yet come through.

29 *YLLS*: 32/3b.

30 See Bray (1995, 1997: 326–334) on the lay and medical obsession with menstrual regularity, and the often-fatal consequences of neglecting amenorrhoea.

31 At the time that Xu Dachun and Ye Gui were practising, the city-dwellers of the Lower Yangzi cities, both women and men, were apparently avid for tonic drugs that would replenish *yin* and were apt to refuse other types of prescription, even though – as Xu Dachun sourly remarked in his essay 'On patients', *Bingjia lun* – they had not the slightest understanding of *yin yang* theory (Chao, 2009: 39–40).

> This is the only reason. We must simply nourish Blood and *qi* and increase the bodily secretions, and the menses will flow on their own.' The others doubted that this would lead to quick results and used the emmenagogues as first intended. But [Xue] said: 'This is not the way to treat this disorder. It is a very violent remedy. It will drastically build up the *yang* Fire, and the *yin* Blood will be affected and will flow wildly; if Spleen and Stomach are affected they will waste.'
>
> The result was that her menstrual Blood wouldn't stop flowing, she couldn't eat or drink, and although Xue was called in he couldn't save her.[32]

A young woman was expected to reach reproductive maturity in her third seven-year cycle, and at this point the failure to conceive or to carry a child to term became a medical problem. In reconstructing what we might call the *ethnobiology* of reproduction in late imperial China, it is important to note the essential ambiguity of some crucial concepts or conditions. I have discussed elsewhere the ambiguities inherent in the practice of menstrual regulation, *tiaojing*.[33] Because menstrual regularity was at the root of female health and fertility whereas blocked menses were believed to lead quickly to debility or even death, women whose menses were delayed felt justified in using drugs to bring on the menses – although it was also recognized that even skilled physicians found it almost impossible to distinguish between the physical characteristics of early pregnancy and blocked menses. In the *fuke* cases, blocked menses, *jingbi*, figure as the cause of life-threatening syndromes, as in the case of a woman of twenty-four *sui* treated by Xu Dachun. She had tried countless remedies to no avail. By the time Xu was called in she was unable to eat and her family were at their wits' end. Over several days he gave her a series of prescriptions to loosen the phlegm, which he had diagnosed as blocking her system; soon a menstruation was brought on which expelled waste matter from her womb; within a month regular, spontaneous menstruation was re-established and the woman regained her health and vigour.[34] The happy ending recorded here is not the birth of a baby: the woman's restoration to health, vitality and menstrual regularity implied to any Chinese reader that her fertility had also been restored.

Other causes of menstrual irregularity or barriers to conception noted in Xu Dachun's cases include vaginal discharges, often triggered by depression or anxiety; fevers; and obesity. Xu records the case of a woman whose womb's membranes were clogged with fat, so that although she was menstruating regularly she was unable to conceive. Xu treated her with 20 or 30 courses of drugs to disperse the

32 *YLLS*: 32/3a–b.

33 Bray (1995, 1997: 326–334).

34 *YLLS*: 32/3b.

fat and open the womb, and the next year she finally gave birth to a child.[35] Knots, *jie*, or accumulations, *juji*, which hampered healthy physiological flows, also caused infertility:[36]

> I met a soldier who said to me: 'When my foolish wife was as yet unmarried she had a cold accumulation under her heart like the cover of a cup. When touched it made a sound like water; if stroked with a hot hand it felt like ice. She has now been married fifteen years and I fear my line will be cut off, so I wish to divorce her.' I stopped him, saying: 'If you use the drugs I give you her illness can be cured and a pregnancy achieved.' So the soldier brought her for consultation . . . Her pulse did not have the cast of infertility, *wuzizhi hou*, and I said she could be pregnant in under a year. The soldier laughed and said his old woman would try it . . . [Prescriptions listed] . . . In less than two months her Blood and *qi* were harmonized, and within not many months she was pregnant twice; both children lived.[37]

Although the sex of the children is not given, their birth signals a satisfactory conclusion of the case, so probably at least one was a boy.

Knots and accumulations were associated with the repression of strong emotions. Body and spirit were weakened when natural circulation was impeded, and this could cause a "ghost pregnancy", *guitai*:[38]

> A woman over thirty, whose periods had stopped for eight or nine months; her abdomen was distended and grew daily, her complexion was sometimes dark, sometimes yellow. She took drugs to test for pregnancy [see below] but without response. I examined her. Her pulse was thready. When her complexion was dark she alternated between hot and cold; the disorder was located in Liver and Gallbladder. When her complexion was yellow her intestines ached dreadfully, she felt tired and resisted eating; the disorder was located in Spleen and Stomach. This was not a true pregnancy. Melancholy had caused a knot, *jie*, damaging Liver and Spleen so the *qi* of Gallbladder and Stomach was not able to process and cleanse properly. *Evil spirits had thus been able to possess her – this is called a ghost pregnancy.* I boiled together the two decoctions *guipi* ['restore Spleen'] and *xiaoyao* [a gentle prescription for building up female vitality], with 3 scruples of 'spirit-beheading cinnabar', *zhanguidan*; she discharged great quantities of dirty Blood and turbid water

35 *YLLS*: 33/11a.
36 Furth (1987); Zeitlin (2007).
37 *YLLS*: 32/10a.
38 Wu (2002); Zeitlin (2007).

> in which was a [kind of] womb, and in this womb was a clot of Blood which closely resembled a ghost's face. *So I have noted this to signal a curious event.*[39]

Here Xu attributes the pseudo-pregnancy to possession by a ghost, rendered possible by melancholy and its physiological effects. Ghost pregnancies were an acknowledged medical phenomenon, sometimes attributed by physicians to supernatural causes (possession by spirits), sometimes explained as a natural if abnormal outcome of emotional or physiological ill health disrupting the processes of pregnancy:[40]

> A woman whose abdomen gradually grew bigger as if she were pregnant. When she reached the tenth month she asked for drugs for an easy birth.[41] I examined her: her spirit, *shen* [one of the body's vital forces], was in difficulty and her pulse was very weak; clearly this was not a fetus in good shape and it would be difficult to give her drugs. Not many days later she indeed gave birth to half a bucketful of white worms. This was because the woman's original *qi* was extremely depleted, so although semen and Blood had congealed[42] she was unable to form a fetus and they formed a rotten, dirty mass; after a while dampness was transformed into heat, and heat and dampness together produced the worms so that she appeared pregnant. This woman died less than a month afterwards.[43]

Here the mass of worms of which the sick woman is delivered is explained by the physician not as a monstrous phenomenon but as the product of a natural transformation of blood and semen under conditions of poor health. Yi-Li Wu reminds us that nothing could be taken for granted about what the womb contained even in the last stages of pregnancy: in a society lacking the certainties implied by chemical pregnancy tests, ultrasound and fetal testing, human reproduction was

39 *YLLS*: 32/22a–b, emphases added.

40 Wu (2002).

41 It was common to try to control the size of the fetus in late pregnancy to avoid difficulties in childbirth. "Thin fetus drugs" based on sweet orange peel are mentioned in *materia medica* as early as the ninth century as a way to avoid difficult childbirth, and were taken from the end of the fifth month. However, the drugs had a tendency to weaken the *qi* energy of both mother and child, so the mother had less strength to deliver and the child would be weak and difficult to raise. As an alternative, physicians recommended "life-sustaining drinks" to increase energy; *Shangyou tang yi'an*: 6/38b.

42 Conception occurred when male and female generative essences (semen and blood) mingled harmoniously and congealed in the womb to form the solid mass that would develop into the fetus.

43 *YLLS*: 32/21a.

experienced as infinitely mutable.[44] Some physicians claimed to be able to diagnose pregnancy by pulse after a month or two, others felt they could not be sure until at least three months had passed. A number of well-known pregnancy tests existed, prescriptions consisting of drugs such as Buddha's hand and *xiaoyao* powder, which nourished and mildly activated the blood.[45] If the woman who used such a test was not pregnant it would stimulate a menstrual flow; if nothing happened, or if there was a movement in the abdomen, the chances were that she was pregnant. A physician might prescribe such a drug to confirm his diagnosis, and a woman could easily obtain it herself from a herbalist or pharmacy. When a woman missed one or two periods, then, she might simply presume she was pregnant, she might procure a mild drug to test for pregnancy, or she might construe her amenorrhoea as the dangerous condition of "blocked menses" and resort to one of the emmenagogues referred to earlier – prescriptions of violent, heating blood-activating drugs, which were well-known to double as abortifacients.

If a woman suspected that she was pregnant but hoped that she was not, the uncertainties of early diagnosis together with the imperative of "regulated menses", *tiaojing*, without which long-term health and fertility were impossible, offered legitimate prescriptions to resolve the issue without even framing it as a step to terminate a possible pregnancy. After all, 'if a spontaneous miscarriage takes place in the first month, nobody realizes that it is a miscarriage – they simply say they were not pregnant; they do not realize that they were already pregnant and have miscarried, *anchan*'.[46] The medical cases only tell us when things went badly wrong, as they did for 'Wang Guo's wife, thirty-five *sui*, who detested giving birth' and mistakenly dosed herself with very powerful abortifacients, provoking a violent haemorrhage; or the woman in her fifth month who dosed herself with a home-made abortifacient of "boiled red pills", *jianhongwan*; or the woman of almost forty with a weak constitution who induced abortion herself. The second case concludes, after a series of subtle prescriptions, with the fetus being gradually calmed, *jian an* (implying that both the patient and the pregnancy were saved); in the first and third cases the fetus is lost but the patients are completely cured, *quan chou yi*.[47]

Abortion was not illegal in late imperial China, although the law punished doctors or midwives who caused a woman's death by prescribing dangerous drugs.[48] Most medical case histories offered no moralizing comments on women who tried to

44 Wu (2002).

45 Bray (1997: 320).

46 Ye Gui, quoted in Huang (1983: 23).

47 *YLLS*: 32/21b; 32/21b; 32/21a–b, taken from the sections on preventing miscarriage, *fang taizi zhui*, and on miscarriage, *taizhui men*.

48 Luk (1977).

end their pregnancies, although they were scathing about physicians who ignorantly or unscrupulously prescribed dangerous blood-moving drugs. Physicians present themselves as giving priority not to the fetus, but to the woman's survival and long-term health. Not surprisingly, however, their recorded cases usually show how their exceptional skills enabled them to save both mother and child. The case collections include long sections on the health problems of pregnant women, and equally long sections on "calming the fetus", *antai*, reducing fetal movement or agitation, preventing miscarriages and ensuring healthy fetal growth. The disorders of pregnancy often pitted the survival of the fetus against that of the mother, and treatment required a carefully planned sequence of therapeutic tactics, as in a case of heat damaging the *yin* of the Lung recorded by Ye Gui:

> First cold then hot, coughing and irrigation: this is a Lung disorder of the spring months and warming wind. The wind constitutes a *yang* pathogen, and the warming is gradually transformed to heating; this alien *qi* which causes an infection is the diurnal, seasonal *qi*. . . . The use [of the bitter, heating drugs normally prescribed for Lung complaints] before the fetus is formed [in the third or fourth month] is forbidden because they produce great heat and agitation, conditions which endanger [the pregnancy]. [In this case] one is recommended to clear away the Lung agitation and to moisten the Lung parchedness. If the fetus is given cooling drugs then it will be calmed, if the disorder is dispelled the patient herself will be calmed – this is what I call non-replenishing replenishing. The physicians of old first cured the repletion [which generally manifests in heat], for the repletion is the pathogen. [Prescription includes a series of drugs for clearing heat.]
>
> Then: Gasping and fever diminished by half, four limbs slightly chilly, uncomfortable feeling in abdomen. There is the worry that the *qi* of the fetus may be affected above. Yesterday she took the clearing and moistening prescription and broke out in beads of sweat: one could see that the bitter and parched were exhausting the Blood which then helped the fever. Today however the agitation and thirst have stopped. *I ask the original cause of the disorder. The damage caused by sorrow and fear* had nourished the Liver *yin*. *The cure* is to stimulate the Kidney secretions, to stabilize and preserve the embryo while regulating the sick body. Prescription for Restoring the Pulse, omitting [the warm drugs and those that dispel cold Wind] and adding [cooling, heat-dispersing drugs from the previous prescription].[49]

49 *LZZN*: 9/35b–36a, emphasis added.

In a vicious circle, fear and anxiety about the risks of pregnancy often made a woman ill. Some cases tell us of a husband pleading with a doctor to give drugs that will expel the fetus in order to save his wife:

> The wife [of a junior official]: her fetus was leaking [slight but continuous bleeding]; doctors had treated her but the leaking did not stop. Her husband, given the circumstances, wished to have her abort and consulted me. I prescribed Buddha's hand powder so that if the fetus could be calmed it would, and if not it would abort, but in a natural fashion.

This case has an unhappy ending: Xu goes away to visit his in-laws, and while he is absent the anxious husband resorts to another physician who administers powerful blood-moving drugs. The unfortunate woman haemorrhages and dies.[50] In the numerous cases of sickness during pregnancy, excessive morning sickness, or threatened miscarriage the author of the case usually records the happy outcome of his ministrations as including the "calming" of the fetus, *antai*; sometimes but not always he also records a successful birth.[51] Sometimes the fetus is dead in the womb, and the happy outcome is a successful diagnosis of this tricky condition, followed by a prescription of drugs to expel the dead fetus and restore the women to health.[52] Occasionally we are given a narrative which shows that the physician has followed his patient over a period of many years, as in the following case treated by Wang Ji and included by Xu Dachun in his section on "Menstrual regulation", *tiaojing*:

> A married woman, twenty-one *sui*: for six months at her period her abdomen hurt as if it was being scraped; the pain was so bad she wanted to die . . . [prescription]. After nine years she had a child, and afterwards she took the medicine several times with success, but after fifteen years the prescription lost its effectiveness and Wang examined her once more . . . [concluding that it was no longer the same disorder]. 'When she first became ill her form was thin and small but now it is fat and large. The classics say that when a thin person has much Blood they suffer from heat [which produces repletion], but when a fat person has much Blood they suffer from depletion [the opposite of repletion], so how could the cure be the same?'[53]

From the standpoint of reconstructing demographic mentalities or reproductive cultures, one might have hoped that a case of this kind would tell us how many

50 *YLLS*: 32/20a–b.

51 *YLLS*: 32/13a and 32/15a are just two of many such cases.

52 For example, the two cases described in *YLLS*: 13b.

53 YLLS: 32/1a.

pregnancies and live births the woman experienced during her long-term treatment for painful menstruation. But the point that Wang Ji is arguing is that constitution changes with age; we cannot tell if his patient had one child or several during the course of his successful management of her dysmenorrhoea. Some cases, like the by one Master Han, included by Xu Dachun in his section on leucorrhoea, give far more detail of a demographic kind:

> A woman in the mountains, over thirty, who had been pregnant eighteen times: nine were premature or stillbirths, seven infants joined their ancestors young . . . The woman's fright and depression were so extreme that she became totally confused and unaware of human affairs. Her mouth, lips and tongue were all ulcerated; sometimes her gullet would become blocked although the lower part was empty, and she suffered from leucorrhea. She was in this state for more than forty days; sometimes she became slightly conscious, desiring to hang herself in unbearable grief at her condition.
>
> A physician gave her a cold course of treatment [a prescription lasting several doses] to cut off the upper heat but then below the symptom of leucorrhea became even more serious. Another gave a hot prescription together with a syrup of aromatic drugs, and then a fever began to mount in her lower body, she became dizzy, and wished for death. I took her pulse and recognized the symptoms of the extinction of *yang* [i.e. she was near to death]. [The family] wept bitterly, saying: 'The ancestors' descent has not yet been established yet they are near to mistakenly killing our elder brother's wife.' I hastily took salt and boiled up 9 scruples of aconite [a powerful and poisonous drug for dispelling inner heat] as the principal drug, compounded with the juice of peppermint [dispels fever] and assisted with ginger ash, asparagus and the like of Schizandra sinensis seeds, simmered in water and then cooled in a well. I administered it, and before she had finished the dose she fell into a heavy snoring slumber and slept through the night, after which she was fully conscious again.
>
> But after some time there still was no heir though two daughters had followed each other in quick succession. She wept for the pain, and all her relatives distant and near grieved. A friend, Zhao Xianchang, was alarmed and asked me, 'Sir, what method do you have for dealing with this?' I said: 'In the books of prescription true must be matched with true and false matched with false. Outside we had a false heat and therefore I followed it with a falsely cold drug; but inside we had a true cold and therefore I countered it with a truly hot drug. Thus inner and outer were brought to harmony and the disorder was resolved. If I treat her continuously with 'female gold elixir pills' then the confusion [of true and false symptoms] will be transformed: not only will her symptoms be dispelled but her original *qi* [vital forces] will

> be regulated.' And in her thirty-seventh year she gave birth to a son who has now inherited.[54]

This woman had been pregnant 18 times by her early thirties (a vastly greater conception rate than I have seen mentioned in any other case); only half the pregnancies resulted in live births and seven of the nine infants died; the two who did survive were daughters. After treatment she gave birth to two more daughters, and eventually, at 37, to a son who lived; since no pregnancy problems are mentioned in the later part of the case, we may assume there were no further miscarriages or infant deaths. So here we have five live children – or, as far as the family was concerned, a single son – at the cost of 21 pregnancies. It was sorrow and desperation about the lack of a son that brought the woman to the edge of death in the first place. The immediate cure was effected with drugs – but it was recognized by all involved that the only final cure, which would restore the woman not only to health but also to happiness, was the birth and survival of a son and heir. This healed both the woman and the family; a happy ending was thus achieved and the case could at last be considered closed.

Concluding remarks

I began this chapter by suggesting that medical case studies might illuminate the demographic mentalities, or reproductive cultures, which demographic historians hope to recover as context for reproductive choices, or even to incorporate into their models. Yet clearly it is no simple task to integrate the quantitative and qualitative approaches.

If more *fuke* cases were as explicit about an individual's complete reproductive history as the last case I described, they might offer real possibilities for relating fecundity to fertility. Much more typical, however, are the narratives that conclude by informing us that 'the fetus was secured and the woman gave birth to two children in a row', or 'a child was born within a year', or 'she gave birth to a succession of sons and daughters'.[55]

What might medical cases reveal about the context in which people made reproductive decisions? The new demographic historians argue that the late imperial period was marked by low marital fertility rates and flat fertility curves characteristic of a society using some form of birth control, most likely selective infanticide. They propose that families preferred to invest in a few children rather than to maximize the number of sons at any price. Medical cases say nothing of infanticide. In other respects, they support demographers' suggestions that late

54 *YLLS*: 32/7a.

55 *YLLS*: 32/21a; 32/10b–11a; 32/8a.

imperial families, or at least the families likely to consult physicians, were not interested in babies, or even boys, at any price. The structuring of the medical narratives shows that many physicians and their clients put the long-term health of the mother before the life of an individual fetus, and that they considered menstrual regulation an indispensable precondition for female health and fertility. We can also infer from *fuke* theory and cases that women not infrequently terminated early pregnancies with drugs that we would term abortifacients, but which in late imperial terms were pregnancy tests. So the medical cases indicate that women had means and rationales for achieving what in our terms would be a degree of control over their own fertility, and show us social contexts in which such control was apparently condoned. Yet they also show us the heavy price many women had to pay in order to produce a socially desirable family. They tell us tales of agitated fetuses; of women exhausted by child-bearing; worn out by miscarriages; driven mad by their failure to produce a son; or dying in childbirth. They describe how sickness can curdle the contents of a woman's womb and turn it into worms. In contrast to the image of conscious decisions and rational planning proposed by the demographers, the medical case histories of late imperial China remind us of the perpetual ambiguities and uncertainties surrounding pregnancy, and the unfathomable mutability attributed to natural reproductive processes.

SECTION III
Androtechnics
The writing brush, the plough and the nature of technical knowledge

This section does not claim to offer a comprehensive account of the technologies associated with masculinity in late imperial China. Rather, it explores the different expressions of male technical expertise and approved masculine identity generated by one heavily symbolic and materially essential domain of activity, namely farming.

If the iconic womanly work across the classes was the making of textiles, farming was the work that men of high and low status could all undertake with pride. These chapters examine the domain of farming and agronomy to investigate how distinctions between different virtuous male personae (farmers, landowners, serving officials and emperors) were produced and expressed. They investigate different aspects of the production of agronomic science, focusing on how material practices were documented and inscribed, how the goals and forms of knowledge were inflected by the social position of the author, and the conditions in which new kinds of knowledge were produced and transmitted. The historical evolution of technical knowledge and of verbal and graphic techniques for organizing this knowledge suggest interesting comparisons with the rise of 'useful and reliable knowledge' and science, and with the forms of masculinity associated with these activities, in Europe. Here it is argued that we can only understand the historical trajectories of agronomy in imperial China if we bear in mind that material and symbolic efficacy were inseparably entwined in the statecraft imperative to 'promote agriculture'.

6

SCIENCE, TECHNIQUE, TECHNOLOGY

Passages between matter and knowledge in imperial Chinese agriculture

When Joseph Needham began publishing *Science and Civilisation in China* some 60 years ago, a routine critique was that he implied misleading connections between science and technology. Needham divided the work into volumes corresponding to the major modern "pure" sciences (mathematics, physics, chemistry, biology), each followed by corresponding applied sciences or technological fields. Magnetism and building, for instance, fell respectively under the categories of applied physics and physical technologies. This strategy for connecting ideas and matter expressed Needham's belief in the reciprocal impact of science and technology, and of scholarly and artisanal knowledge in the production of science.[1] At another level of communication, it also reflected the everyday connections people draw between "pure" and "applied" scientific knowledge, and this was certainly an important factor in the immense public success of Needham's project. Within the discipline of history of science, however, critics such as Lynn White raised strenuous objections, arguing that by adopting a framework that designated technological activities as applied forms of science, Needham used pre-modern China's undoubted technical brilliance to imply an exaggerated level of scientific achievement. Meanwhile, in his influential collection of essays entitled *Clerks and Craftsmen in China and the West*, after arguing for the fertile contributions of official-craftsmen interactions in the early phases of Chinese history, Needham suggested that in the long run Confucianism and the bureaucratic concerns of the late imperial state were inimical to the pursuit of science and technology, not least because the neo-Confucians who came to dominate the late imperial governing elite held artisans

1 Here Needham acknowledged Edgar Zilsel as an important influence (see Needham, 2000).

and their working knowledge in low esteem. In other words, Needham was inclined to argue for a chasm between the academic knowledge valued by the late imperial state, and the craft skills that built the material world it governed.[2]

This position has since been amply refuted by an impressive body of research charting the relations between rule, knowledge and material action in late imperial China. What constituted significant or crucial domains of natural knowledge in China did not always conform to Western patterns. Nevertheless, as recent scholarship has amply demonstrated, the late imperial Chinese state and its officers played key roles in producing and disseminating new forms of scientific knowledge, and in mobilizing technical instruments or practices to generate such knowledge, particularly in domains of such obvious statecraft relevance as astronomy or hydraulics.[3] Following the strictures of the new, critical history of Western science, however, and still recoiling from Needham's supposedly imprudent and profligate use of these terms, except when referring to imports from the West, historians of China often prefer to speak of knowledge and practice rather than science and technology.

In this chapter, I argue for the value of reinstating the terms science, techniques and technology as tools for a more precise analysis of governmentality and the workings of power in the late imperial Chinese state. My tactic is to use these three categories explicitly, in order to highlight flows between matter and ideas in the production and reproduction of knowledge. To this end I propose the following definitions.[4] *Science* I define as knowledge about natural, material processes expressed in declarative, transmissible form; its representations generally aspire to be authoritative beyond the time and place of their production. *Techniques* I define as the skilled practices that go into the material production of knowledge as well as the production of artefacts. *Technology* denotes social-material networks or systems, including sets of techniques and equipment, but also trained personnel, raw materials, ideas and institutions. Technology in this sense of networks or systems is here of two-fold interest, first as generating material goods and social relationships, and second as contributing to the production or reproduction of the kind of knowledge I have termed science. Thus defined, science, techniques and technology are not separate kinds of activity but rather overlapping phases of an organic process of knowledge production.

The case I will discuss is agriculture. Farming is a material activity that spans the gap between pre-modern societies and our contemporary world. It is very obviously a materially rooted domain of knowledge, embedded in social networks.

2 White (1984); Needham (1969).

3 For example, Elman (2005); Cho-ying Li (2010); Schäfer (2012).

4 Like Malinowski's (1948 [1925]) reformulation of the distinctions between magic, religion and science, I would argue that thus defined, science, techniques and technology constitute three interlocking categories, that can be found in any human society.

It is furthermore a domain of technical knowledge and practice in which the Chinese state has actively intervened since its inception: recording and transmitting technical information, investing in infrastructure and improvements, and directing local practices.

Modern agriculture is generally thought of not as a science in itself, but rather as a science-based productive activity (in today's popular terms a technology) that applies knowledge from several scientific fields. Following this convention, in *Science and Civilisation* agriculture, *nong*, was classified as a "biological technology".[5] But if as historians we take a science as a domain of natural knowledge considered primary in its own time, then in imperial China agriculture was a science in its own right. Imperial China was, throughout its duration, an agrarian empire in which the state levied taxes and landlords demanded rent from the peasantry; the land was the source of sustenance and of wealth, of production and extraction. Agriculture was not only the material but also the social and ethical basis of the polity, or – as Chinese political philosophers put it – the "root" or "foundation", *ben*. In consequence, it was the object of systematic enquiry and intervention by the governing class.

What kinds of knowledge – and whose knowledge – were inscribed in the Chinese agricultural texts, which have come down to us in such abundance, and in the farming landscapes whose distinctive features we can track over two millennia? Agriculture is a domain of knowledge where it is impossible to separate ideas *about* matter from the struggles *with* matter that generate them. Nor are the knowledge and techniques politically neutral: the norms of good farming are a very powerful instrument for ordering society. Agriculture, then, constitutes a wonderfully rich case of the intertwining of knowledge, practice and power; of the production of ideas, material goods and social relations. How is this reflected in the agricultural texts that circulated among the ruling Chinese elite and filtered down to peasant farmers through magistrates and landlords? I address the formation and the power of agricultural knowledge in imperial Chinese society explicitly in terms of the relations between science, techniques and technology. I offer selected examples of the material access to knowledge characteristic of the different groups of actors, and the various processes of their translation or encoding into textual or graphic inscriptions, including, for instance, how the three-dimensional farm implements of peasants and their uses were translated into two-dimensional printable form. I also address the question of how "mobile" and how "hard" the facts of Chinese agronomic science proved, given a written tradition that developed over more than 2,000 years, and dealt – among other variables – with a huge territory including deserts and tropical forests, dry plains and irrigated rice lands.

5 See the title page of the volume on agriculture in *Science and Civilisation in China* (Bray, 1984: v).

Before plunging into the Chinese materials, I begin with a general discussion of the relations between ideas and matter, knowledge and power, in terms suggested by recent science and technology studies. Here I take modern agriculture as the illustration, because I feel it is helpful to argue through familiar facts the case for the rather special definitions of science, technology and technique that I have just proposed. In this section I focus on how knowledge flows between science, technology and technique, its passages from matter to idea and back. I discuss processes of inscription and what Feenberg calls the "codes" embodied in such inscriptions, and I draw on Latour's concepts of "immutable mobiles" and "hardening the facts" to address connections between technology and empire.[6]

Science, technique and technology

In the framework of *Science and Civilisation* Needham divided domains of natural knowledge into the knowledge – or science – itself, and its applied forms – the associated technologies. This approach reflects what Pfaffenberger calls the "standard view" of science and technology,[7] a model of the relations between ideas and matter that became conventional with the parallel rise of academic scientific disciplines and of engineering and industry in the nineteenth century. Although dismissed by today's historians of science as a discredited caricature, versions of the "standard view" still regularly surface among scientists, educators and the general public, and despite its problems this model of knowledge flow offers some illuminating insights into the material production of formalized human knowledge.

The "standard view" makes a categoric distinction between science and technology. Science offers new understanding about the nature of matter: it quarries the messy material stuff that surrounds us; isolates, identifies, purifies and studies nuggets of it; then dematerializes the natural stuff into elegant ideas, of universal validity, expressed in verbal or numerical form. Technology takes the universal knowledge generated by science and applies it systematically to specific transformations of matter, thus rendering the world in which we live more orderly, productive and convenient. Knowledge is produced by scientists and flows in one direction, from science to technology, from the laboratory to the factory.

This model of the production and application of ideas about matter left little or no place for the kind of embodied knowledge characteristic of preindustrial production, what we might refer to as "craft" or "skill", knowledge embodied in *techniques*. Techniques are customary procedures for the transformation of matter that are routinized yet flexible, for they have to cope with the variations typical of

6 Feenberg (1999); Latour (1987).

7 Pfaffenberger (1992: 493–495).

8 For example, Lemonnier (1993).

natural materials;[8] furthermore, techniques are notoriously resistant to translation into either words or numbers. The technology of industrial production was understood as epistemologically distinct from artisanal techniques in two senses. First, it involved the conscious application of previously formalized scientific knowledge to material processes. Second, the logic of industrial design was to eliminate the material irregularities that were the true test of a craft-worker's skills, while building as much as possible of the regularized processual core of a productive process into the machinery itself. The production of scientific knowledge, meanwhile, was understood as essentially a disembodied process – the mind was crucial, the hand only incidental, even the hand of the scientist conducting the experiment.

Recent scholarship has brought techniques back into both science and technology, looking from a variety of perspectives at how ideas and matter connect through concrete practices. Historians, anthropologists and sociologists of science now routinely treat sciences – at least sciences as they have emerged in the post-Renaissance West – as materially rooted practices in which the materiality both of the object of study, and of its observation, are intrinsic to the generation of knowledge. Techniques and skills are re-established as an essential element in the generation of scientific ideas, a key dimension of experiencing, perceiving and knowing matter. Parallel arguments are made for technology. Standardized procedures have to be adjusted through skill and technique to adapt to the roughness and intransigence of real matter in its local context, and the questions thus raised may generate new scientific research programmes.[9] In other words, science is not just knowledge *about* matter: it is also knowledge that comes *through* matter. Techniques mediate dialectically between knowledge and matter, and as such they are essential both to scientific investigation and to technological practice, underlining the epistemological continuities between the two domains.[10]

Another point that recent history and sociology of science has effectively hammered home is that science is not the private activity of lone geniuses, but a communicative activity. Sociologists of technology distinguish between invention and innovation: devising a new solution to a material problem in your garden shed is very gratifying, but only if the solution is taken up by society at large will your thrilling new device step into history. Similarly, the impact of a new fact or theory in science depends not simply on its accuracy but on how effective it is first in persuading fellow scientists, and then in translating into ideas or institutions that affect society at large. Hence the recent emphasis on inscription, on social networks, and on the translation of science into what Foucault has referred to as governmentality and technologies of power.

9 For example, Staudenmaier (1985: 103 ff.); Akrich (1993).

10 Felicitously described in a recent collection on the history of inquiry and invention as a 'complicity between contemplation and manipulation' (Roberts *et al.*, 2007: ix).

To be validated, science has to *persuade* people. To function as science, that is as knowledge legitimated within a scientific community, knowledge must be expressed in declarative form. It is at that level that it is tested, contested, validated, and fitted into a system, and so acquires power as an authoritative affirmation about the natural world. This applies equally to science today and in pre-modern societies, and it is the reason why I have chosen in this chapter to define science as textually or graphically inscribed knowledge. In the process of inscription, the knowledge is often further encoded by translating it into a specialist technical vocabulary, verbal or visual. The persuasive power of a scientific fact or theory is closely related to its ability to travel through time and space, and to subsume heterogeneity. Hence the drive in inscription to simplify, to homogenize – to produce representations of the nature of material processes that are accepted as generally valid beyond the time and place of their production. Latour calls such representations "immutable mobiles", explanations or artefacts that can be transferred through space or time without requiring modification; as facts are "hardened" into simpler and more mobile forms, more and more of the uneven contours of raw matter in its native environment are flattened out into manageable, regular elements.[11]

Part of the power of a scientific idea lies in its mobility, and in the simplifications it is able to effect in its inscriptions. Part lies in its capacity to produce thick, dense, deep-rooted networks that ensure its stability. Scientists persuade each other, and eventually the broader society within which they operate, not only by mustering plausible arguments, but also by enrolling dense and durable networks of human, institutional and material supporters. The "standard view" tells us that technology "applies" scientific knowledge. But we can take this further to argue that industrial and other technologies reshape the material world in ways that confirm the truth of the corresponding science. Technology rematerializes the abstract universals of science, reshaping our surroundings into new configurations almost as elegant in their simplicity as the equations that they embody. Messy natural stuff is processed into homogeneous materials: pure chemical compounds, precise alloys, wire of exact gauge; in other words the neat pieces that allow processes of mass production to function identically anywhere in the world, generating components or commodities of regular, geometrical shape and standardized size. Furthermore, while the "standard view" represents knowledge as flowing out from science and into technology, recent studies have underlined how the constellation of social institutions, material artefacts and technical skills and practices – representing often huge financial, educational and political investments – which validates a set of scientific ideas often in fact directs the further development of that knowledge.[12]

11 Latour (1986: 17).

12 For example, Traweek (1996).

Here I come to what Feenberg calls a "technological code". A technology is a material system in which tacit understanding and explicit science both play a role in naturalizing social hierarchies and encoding cultural values. Feenberg gives the example of factory machinery in early nineteenth-century Britain built to a scale suitable for child operators, thus encoding the social and legal endorsement of child labour.[13] Another good example is the calculation of farming efficiency. Everyone knows that agriculture produces corn, and one easy way to measure the efficiency of a particular agricultural technology is to measure how much corn per acre is produced. As a simple ratio, "efficiency" thus calculated appears a natural, neutral way of evaluating one system in comparison to another: Iowa farmers are more efficient corn-growers than Mexican peasants. Yet once other factors are admitted into the analysis (levels of pollution or of consumption of fossil fuel; biodiversity; optimal farm size; numbers of people employed or gaining a livelihood; subsidies; national export policies; impact on farmers in developing economies), the political dimensions of the technology, and of the measures chosen to evaluate its performance, become plain.

Agriculture is a technology where it is usually rather easy to identify the social or political interests that lie behind what is defined as "good practice" or "good science". "Scientific" farming first emerged in Britain in the late eighteenth century as an activity for gentlemen landowners, their factors, and their educated and relatively prosperous tenants. The new methods of draining and consolidating fields, rotating crops and carefully controlling the breeding of livestock required large holdings and access to capital – it was a system of knowledge and practice that smallholders could not afford to adopt, nor could they effectively compete against it. The new science of farming embodied a capitalist logic of economies of scale in the pursuit of greater output and profits. By the mid-nineteenth century a range of industrially produced farm machines had been designed to substitute for human labour, allowing the smooth translation of the model to the vast, sparsely populated plains of the New World where grain was produced for world markets. By the end of the nineteenth century, industrial agriculture had established itself as the unchallenged model of best farming practice, supported by a worldwide network of railroads and steam ships, telegraph cables, factories producing farm machinery and fertilizers, commercial seed companies, land-grant and agricultural colleges, experimental stations and scientific societies. While the early forms of British scientific farming had encoded gentry farmers as their primary beneficiaries, since the mid-nineteenth century cutting-edge agricultural science, even in state institutions, has usually encoded the right of industry and commerce to extract profits from the farming sector.[14]

13 Feenberg (1999: 85).

14 Kloppenburg (1988); Bray (2003); Harwood (2005).

Just how far across time and space can "good science" and "efficient technology" spread their claims? Latour's concept of "immutable mobiles" offers a useful way to think about how science and technology can exercise power over people, and about the challenges or obstacles which knowledge that claims general applicability may face when it attempts to conquer new territory. Modern agronomy produces inscriptions (both textual and material) that are widely believed to be universally valid. Modern farming practices draw upon a network of sciences, among them organic chemistry, genetics, soil science, and – increasingly – molecular biology. Scientific knowledge is produced in laboratories and experimental plots and then translated into technical artefacts (hybrid seeds, reaper-binders, chemical fertilizers, irrigation systems, crop rotations) which farmers around the world are taught to use. The institutions that today support the intensification and further penetration of this normal science of farming include agricultural colleges, global trading networks, international funding agencies, national governments pursuing development, and the research of scientists employed by public institutions or – increasingly – by agribusiness corporations.

The science and technologies of modern agronomy claim general validity. Yet the mobility of normal agricultural science's inscriptions, textual and material, is still impressive when we consider how vulnerable farming procedures are to local variations in soil or weather – to cite just two significant variables. In any application outside the lab the general principles of agricultural science have to be accompanied by instructions on how to adjust them to variations in local context; the inscriptions must be to some extent flexible. But at the same time, the farming methods generated by the inscriptions alter the environment in such a way as to reduce the need for their own modification. Modern agriculture offers a particularly dramatic but by no means unique example of how a technology works to create the conditions in which its science will be valid, and its techniques effective.

One common characteristic of technical artefacts is to extend or multiply human capabilities: a plough as an extension of the human arm, a calculator as one of the brain. Another, less often remarked upon but equally significant, is to flatten out the irregularities of the world around us: buildings are warmer in winter and cooler in summer than the open air; agriculture (together with food storage and processing techniques) smooths out seasonal fluctuations of supply or the impact of disasters. In the industrial era, technology (in the sense both of a set of technical procedures and of a code or ideology) has notoriously been a flattener, homogenizing materials, processes and products around the world. Normal agriculture has moved steadily forward to greater scale and greater flattening power. Quite literally, where bulldozers and lasers level ever-larger fields for convenience of irrigation equipment and large-scale machinery. More metaphorically, where huge areas are devoted to a single crop (monoculture), and often to a single scientifically bred variety of a single crop. Irregularities of time are minimized. Every corn field is sown with the same seed, and needs spraying or watering or harvesting exactly so many days after sowing; genetic modification now permits year-round cultivation and harvesting of strawberries. Space too is homogenized: mountainsides are levelled, swamps drained, and deserts irrigated until they bloom. This material technology is not unlimited in its powers to standardize local landscapes and climate: it is still more efficient, in financial and energetic terms, to grow corn in Iowa than in Alaska, but it would not be technically impossible to turn Alaska into corn farms, at least short-term. The material technologies of industrial farming do more than simply apply the facts of normal agricultural science, they help turn them into harder facts by testing them against initially adverse conditions. And they give enormous mobility to their immutables (general principles of tillage, hybrid seeds) by transforming unfavourable local conditions into replicas, often on a giant scale, of the experimental station.

But if we look at normal scientific agriculture as a sociotechnical system, we see that its mobile immutables move more easily through some social networks than others. Like the scientific farming of the eighteenth century, the scientific facts and artefacts of the contemporary system encode productivist economies of scale and therefore work better for large-scale farmers with access both to land and capital; they have a tendency to eliminate smallholders. In nations such as the United States, Canada or Britain, where normal agriculture took hold early, the weakness of smaller farms trying to operate within or as an alternative to this system is construed as inefficiency. In countries such as India or France, many small farmers soon learned to discern and challenge the code embodied in such apparently neutral artefacts as hybrid or genetically modified seed, and public support for small farmers claiming the political right to survive has obliged governments to moderate their initial enthusiasm for the new technologies.[15]

In other words, what is considered an efficient or a poor technique for sowing rice, a good or poor scientific rationale for the breeding of cattle depends to a significant extent upon the social and cultural context. Within a society that largely shares values and institutions across class and interest groups, the hardness of facts, the transportability of the immutables, may be tested principally against local variations in natural conditions. But if the central state attempts to impose settled farming on swidden farmers, if landowners want to switch to sheep and throw out their tenants, or if tenant farmers adopt a new cropping system that undermines traditional rental arrangements – then these proposed transformations will meet fierce resistance. If the resistance is overcome and a new social contract is achieved, then agricultural science and technology will serve as powerful instruments for maintaining the new social order and for inculcating the subjectivities associated

15 Bray (2003).

with that order. It can be argued that even in modern nations where it occupies only a small portion of the population and of national economic production, agriculture is still a fundamental instrument of governmentality and encoder of shared or contested social values.[16] In an agrarian state such as imperial China, the symbolic and material importance of technical choices in agriculture were a constant preoccupation of the ruling elite.

Agronomy as state science in late imperial China

In the modern world we take it for granted that governments play a key role in promoting agricultural science and development, but this is a rather recent development. Although almost all states before the nineteenth century were built upon agrarian economies, few put the technical promotion of agriculture at the centre of their philosophy of rule. Most governments of Western states, for example, played little direct role in the production or dissemination of agricultural knowledge before the nineteenth century.[17] Imperial China, however, was from its inception an agrarian state in the strong sense of the term. The production and circulation of agronomic knowledge by the state was a key technique of government.

What kind of "agronomic science" did the imperial Chinese state produce? A great deal of the technical knowledge that the state produced and circulated under the heading of *nong*, agriculture, consisted of encodings of the material knowledge and skills of practising farmers in a precise specialist language developed over centuries.[18] Certain types or forms, however, depended on the special resources available to officials, or reflected purely official concerns. Here I ask what official strategies for producing agronomic knowledge added to the technical knowledge of farmers. How did they select, encode and deploy knowledge, for what ends, and how successfully could such encodings be decoded and rematerialized as techniques or artefacts? Were official inscriptions simply records of existing knowledge, or did official agronomists also create new forms of knowledge?

I will discuss these questions in relation to the evolution of agronomic knowledge during the late imperial period, focusing on three outstanding examples of agronomic treatises, *nongshu*, written by officials in the period between 1250 and 1650. To highlight their specific characteristics as state science, I compare their strategies for producing knowledge to those deployed in treatises written by private landowners during roughly the same period (a genre of knowledge production that

16 For example, Schlosser (2001) on the contemporary USA, and Bray (2003) or Heller (2007) on France.

17 The United States was a pioneer, setting up an embryonic Department of Agriculture in 1836 as a branch, significantly, of the Office of the Commissioner of Patents.

18 On the evolution of this technical language, its vocabulary and syntax, see Bray (2001a, 2001b).

I analyze more fully in Chapter 7). I begin with a brief sketch of the significance of agronomic knowledge in Chinese statecraft, and some indications of how the relations between knowledge, material practices and matter were expressed and conceived.

At the core of the Chinese tradition of statecraft, *jingji*,[19] was the recognition that *nong* (a term that simultaneously denoted farmers, farming practices, agricultural knowledge and the broad institutional category of agriculture) was the foundation, *ben*, of the human world, the fundamental occupation, *benye*, supporting the polity. An enduring agrarian ideal of the state represented the emperor and his officials as directly responsible for the well-being of a society formed essentially of smallholding peasant households, whose taxes (nominally a modest tenth of their production of cereals and cloth except in times of extreme crisis) were the primary source of state revenues.[20] Farming was the foundation not simply of the economy, but of the social, moral and cosmic order of the imperial state. In late imperial China an iconic landscape of irrigated rice fields dotted with plantations of mulberry trees for feeding silkworms represented not only a source of material abundance, but also a setting in which humans acted out in microcosm the essential interflowing of *yin* (female) and *yang* (male) energies that sustained cosmic and political harmony.[21]

It was thus the duty of the emperor and his officials actively to promote agriculture, *quannong*, for reasons that were pragmatic (to increase state revenues and maintain social stability) but also cosmological and moral: to fulfil the Mandate of Heaven a virtuous ruler must devote himself to "the welfare of the people", *limin*, striving to improve their livelihood and to protect them from poverty and famine, from the risks of commerce and from exploitation by powerful landowners.[22] Official instruction in effective farming methods helped peasant households to fulfil their part of the social contract, feeding and clothing both their families and the state. Proper farming practices also promoted – and were in turn enhanced by – the moral and social values of diligence, propriety and respect fitting to true subjects of the emperor.[23]

Quannong remained an active concern of statecraft throughout the two millennia of imperial rule, involving ritual as well as practical elements.[24] According to the

19 From the classical expression *jingshi jimin*, to manage the realm and aid the people.

20 Wong (1997: 90).

21 See Chapter 2.

22 For a conspicuous example of how an individual official could act to protect the interests of the peasantry, see Rowe's account of the eighteenth-century statesman Chen Hongmou (Rowe, 2001).

23 See, for example, the section on 'Respecting rank and strengthening agriculture' (*Xiaodi litian pian*) in the introductory chapter to Wang Zhen's *Nongshu* (Agronomic treatise) of 1313; *WZNS*: 17–19). Chapters 2 and 3 address how Chinese farming and weaving techniques were deployed as tools for assimilating non-Han peoples.

24 I discuss some aspects of the intertwining of ritual or symbolic with materialist strategies in Chapters 2 and 8.

cosmological doctrine of the Three Powers, *sancai*, humans, *ren*, occupy an intermediate position between Heaven, *tian*, and earth, *di*. Farming, *nong*, requires the application of human action, *rengong*, to draw benefits from the earth, *dili*, in accordance with the seasons of Heaven, *tianshi*. In a given environmental setting it was appropriate human action – high-level technical skills informed by good understanding of natural principles – that made the difference between good or poor harvests and between strong or weak states. Official policy, therefore, focused on developing *rengong*, at the level of farmers' skills, but also at that of local administration, provision of infrastructure, and national policy.[25] It is on the first level, of practical farming skills, that I shall focus here.

Chinese emperors, officials and state institutions all actively participated in the production, dissemination and application of agronomic knowledge on behalf of "the people", *min*, or "farmers", *nongmin*, a social category construed as peasants: simple, vulnerable and uneducated folk with few means and limited experience for whom it was the duty of officials to provide instruction and support.[26] The Ministry of Agriculture, *nongbu*, promoted water control and land reclamation projects; collected and stored grain for famine relief or for price control; set up farming colonies for soldiers or refugees; offered tax relief, loans, seed and basic equipment to the poor; and circulated information about improved farming methods through the network of local magistrates.[27] The official production of texts played a key role in circulating agronomic knowledge; they included technical pamphlets;[28] records kept at district, provincial or national level; imperial edicts and memorials to the throne; and agronomic treatises, *nongshu*. It is on the treatises that I focus in the following discussion, and for further clarity in such a broad field, I take innovations in cotton production as the central theme.

Agronomic treatises, *nongshu*, contain systematic observations and analyses of natural phenomena presented as authoritative knowledge (in the terms of the first part of this paper, *science*). They tell the reader how to apply this knowledge by translating it into forms of material action (*techniques*) that will draw benefits from the earth in the form of good harvests. As a medium of communication, *nongshu* encode scientific and technical information on paper, in technical verbal or graphic formulations that the reader is presumed able to decode, rematerialize and apply

25 See below and Chapter 7 on Xu Guangqi. On the concept of *rengong* see the passage from an essay by Lu Shiyi (1611–1672) on intensive farming techniques, quoted in Li Bozhong (1998a: 448). See also Schäfer (2011a) for an extensive discussion of late imperial theories of the relations between cosmos and human action.

26 In fact the recipients of, or partners in, these outreach efforts also included landlords, much of whose land was worked by tenant farmers (e.g. Shiba, 1998: 159; Li, 2012).

27 For example, Elvin (1973), Golas (1980), Perdue (1987) and Li Bozhong (1998b).

28 A famous example is the instructions for growing quick-ripening Champa rice, introduced into the Yangzi Delta on imperial order in 1012; see Chapter 6.

successfully. In all *nongshu*, whether composed by officials or landowners, techniques (*fang*, technical knowledge, *fa*, specialist methods, or *shu*, techniques or procedures) for cultivating and processing key crops are the centre of attention, along with farming tools or equipment, *nongju* or *nongqi*. However, the two genres of treatise, official and private, approach these techniques in different ways, and at the level of *technology* (the techniques as part of a material-social system) and of *science* (the ways in which natural knowledge is produced and the claims that are made for it) there are notable distinctions between them.

The goal of official *nongshu* was to produce comprehensive, mobile knowledge that could successfully be transferred, through the medium of print, across the vast spaces of the empire and translated into local action.[29] The knowledge they contained was directed at fellow officials whose duties included raising the standards of farming, and of the rural economy more generally, in the area for which they were responsible.

Unlike landowner *nongshu*, discussed in Chapter 7, which presented the *techniques* of farming in the context of a *technology* of farm management, official *nongshu* embedded farming techniques in a matrix of government administration and statecraft concerns. They opened with sections on what we might call the ideology and cosmology of agrarian empire (see above). As well as sections on crops and cultivation techniques, they contained technical chapters on methods of land reclamation, *kentian*, and field systems, *tianzhi* (how to construct various types of irrigated field, polders, terraces, and so on); water control and hydraulic projects, *shuili*; the construction and management of public granaries; and famine foods – all routine concerns for officials. The technical information on farming methods was not intended to teach magistrates how to build a harrow or pinch out cotton themselves: it was a resource for educating and training local farmers. The decoding, the translation of knowledge back into material artefacts and practices and their dissemination, often required the enrolment of such mediators as "instructor farmers" or "skilled carpenters" (see below). Although official agronomists recognized the pedagogical utility of monthly farming calendars, *yueling*,[30] and struggled to devise universal calendars that could be adapted according to simple principles to work in any climatic zone,[31] the overall arrangement of the *nongshu* was not by month but by self-contained topic. A chapter on rice began with a discussion of varieties and their characteristics, then proceeded systematically through seed

29 By the tenth century, woodblock printing was well established as an effective medium of inscription, transmission and archiving for official purposes. Both commercial and official printing and publishing houses played a prominent role in the production and circulation of agronomic knowledge starting in the Song (Chia, 2002; Hu, 1985).

30 A favoured template for organizing private *nongshu*; see Chapter 7.

31 See, for example, *WZNS*: 6–12, *NZQS*: 225–253.

selection, sowing and transplanting through to harvesting and storage requirements; the sequence was universal, the reader decided on the local timing.

One strategy which official *nongshu* adopted for producing knowledge that they hoped could be successfully transplanted in new ground was to identify and document innovations or best practices from around the empire, urging their adoption in less advanced regions (an early form of technology transfer). A second was to provide a spectrum of experience, drawn from history, from different regions or from contrasting environments, from which the reader could select the knowledge or techniques most appropriate to local circumstances. This method worked best if the author structured the materials he quoted, using comparisons to deduce general principles that allowed the adaptation of local techniques, and offering critical analysis based on his own experience.

The basic principles and methods of *quannong* faced many challenges and changes during the two millennia of imperial rule. Between 1100 and 1800 the typical intensive wet-rice landscape of the Yangzi delta progressively colonized most of South China's hinterlands and backwaters, and spread from river plains into wetlands and up into the mountains.[32] In peaceful periods the population grew and so did pressure on land; in response, officials promoted the adoption of new crops or more intensive farming methods, and sponsored the opening of new types of land, and migration to sparsely populated regions.[33] Floods, droughts and crop failures routinely put public granaries and transport networks to the test. After rebellions, invasions or dynastic transitions that devastated vast regions, reducing fertile farmlands to waste and decimating the population, governments would hasten to organize resettlement, providing land titles, seed, oxen, loans and technical instruction. Often, as the following examples suggest, it was some form of crisis that inspired the most innovative approaches in official *nongshu*.[34]

The northern provinces of China had been ravaged by decades of warfare when the Yuan (Mongol) emperor Kublai established a Board of Agriculture, *Da sinong si*, for his new dynasty in 1271 and commissioned a practical handbook to improve farming practices throughout the Yuan kingdom.[35] In 1273 the Board Directors, Meng Qi, Miao Haoqian and Chang Shiwen, presented to the throne a treatise entitled *Nongsang jiyao* (Compiled essentials of agriculture and sericulture), which was immediately printed and distributed all over the country. The most important

32 See Chapter 2 and Elvin and Liu (1998).

33 Often farmers would take these steps on their own initiative, but almost invariably the state would play an active supporting role.

34 As well as the examples discussed here, see also Chapters 2 and 8, which discuss the significance of the scroll-painting *Gengzhi tu* (Pictures of tilling and weaving) as a response to the loss of the northern provinces by the Song government in 1126.

35 The southern Chinese provinces were not incorporated until the Yuan defeat of the Southern Song dynasty in 1279.

new information contained in the work was a 'recent, improved method of cotton cultivation', *jin tian zai mumian fa*.[36]

Cotton cloth had been imported to China as a luxury for centuries, but hemp was still the commoners' alternative to silk. Cotton-processing techniques were highly developed in Central Asia and China's southern neighbours such as Hainan, but the crop was barely known to Chinese farmers at the time. The much-travelled Mongols were quite familiar with the advantages of cotton, which was soft, light and hard-wearing, warm in winter and cool in summer. Furthermore, levels of silk and hemp production had been severely affected by the wars of invasion, leaving the state short of textiles needed to pay and clothe civil servants and the army.[37]

The *Nongsang jiyao* describes how to select the right type of soil; construct raised beds; pregerminate the seed; irrigate the beds; pinch out the seedlings; and so on right through to de-seeding the bolls (the cotton gin was unknown in China until around 1300) and spinning the yarn. Wang Zhen, in his *Nongshu* (Agricultural treatise) completed in 1313, says the government disseminated these techniques in order to introduce cotton where it was not yet known, and to raise standards where it was already cultivated. In 1289 the Yuan government established cotton promotion bureaux in the newly acquired southern provinces. Wang names several regions that had since taken up cotton, noting that in many places it did as well as any native crop, and expressing the hope that it would continue to spread in both north and south.[38]

The *Nongsang jiyao* provides one single method, a general formula, for growing cotton, offering precise details, no variants and no explanations. This instruction-leaflet approach is what one might expect for a new crop untried through most of the areas in which the government planned to introduce it. In these circumstances the generalized *technical* knowledge contained in the written text seems to have allowed for very successful decoding around the country. In terms of broader *technological* context, one powerful incentive for farmers to adopt cotton was that in 1296 cotton was incorporated into the tax system at very favourable rates compared to silk or hemp.[39] We should not forget, however, that there was also a general shortage of cloth, and merchants and markets may have been just as important as the state in promoting cotton (see below).

It is perhaps not surprising that Wang Zhen's *Nongshu*, completed only 40 years after the *Nongsang jiyao*, has nothing to add on cultivation methods.[40] What Wang

36 *NSJY* Chapter 2 as quoted in *WZNS*: 160–168. The context suggests that this method was introduced from Central Asia.

37 Chao (1977).

38 *WZNS*: 161.

39 Bray (1984: 539).

40 *WZNS*: 61.

Zhen does contribute, and this was of enormous importance, is technical plans for the processing equipment (bows, gins, spinning wheels, and so on) recently introduced from China's southern neighbours. Cotton is a short-staple fibre, and it was not easy to process using the equipment developed for China's indigenous long-fibre materials, silk and hemp. The new devices were beginning to revolutionize cotton production in Songjiang prefecture (the region around Shanghai), but Wang noted that they were still little known elsewhere and should be introduced more widely.[41]

Wang Zhen's *Nongshu* was also a response to crisis. Wang was a native of the Shandong province, in the north, and had spent many years as an official in Anhui and Jiangxi, in the south. Most farming regions in China were still suffering from the aftermath of decades of warfare between the Yuan and Southern Song states. Wang urged magistrates to acquire a thorough practical understanding of the best agricultural methods currently available, in order to instruct the peasants under their jurisdiction. His treatise provided informed and accurate documentation of effective tools and methods from around China; he was especially interested in technology transfer between north and south.

Wang Zhen's treatise follows the basic method of documenting existing high-quality practices in order to facilitate their dissemination; in other words, he is not creating new knowledge, but facilitating its circulation. Most of the sections on cultivation methods consist principally of quotations from earlier works. Perhaps Wang's most important contribution here was to incorporate the detailed instructions for the sophisticated techniques of wet-rice cultivation set out by Chen Fu in 1149 (see Chapter 2), thus making them easily available throughout China. These intensive methods gradually spread to the provinces of the middle Yangzi and the far south, continuing to offer useful guidance in less-developed regions long after they were written.[42]

Wang Zhen's truly innovative contribution to agronomy, however, was his development of a new method of encoding technical information, pairing text with illustrations and diagrams in a kind of blueprint (see Chapter 8). The third section of Wang Zhen's *Nongshu*, the *Nongqi tupu* (Illustrated register of agricultural implements), pairs illustrations and text in a systematic inventory of field types, farming implements and equipment, from water mills to sandals, from harrows to machines for reeling silk. The text lists the structural elements of a piece of equipment, giving materials, dimensions, and how they are connected; the illustration provides an overall view of the complete structure.

41 *WZNS*: 416. According to a local scholar, Tao Zongyi, writing in 1366, a woman from Hainan named Huang transformed the Songjiang economy by introducing these devices in around 1300 (quoted in *NZQS*: 968). See Kuhn (1988) for details concerning the development of cotton-processing equipment.

42 Although important advances were made subsequently in fertilizing techniques, crop-breeding and water control, much of Chen Fu's technical advice remained valid.

According to Wang Zhen's friend and fellow prefect Dai Biaoyuan, as well as interviewing the local farmers about their methods Wang 'also made drawings, *tuhua*, of all the varieties of hoes, drills, harrows, and other miscellaneous implements, and had the common people make them, *shi min wei zhi*)', a method which initially caused merriment among his colleagues and bewilderment among the farmers, but which, Dai claims, showed its value in disseminating best practices in just a few years.[43] In other words, Wang was at pains not only to encode the knowledge of these implements in detail, but also to test whether the inscription could be decoded and rematerialized successfully. One essential partner in this exercise of decoding was a carpenter, *mujiang*, whose familiarity with routine solutions to complex mechanical problems Wang assumed would fill in any technical gaps in his encoding.[44]

Wang's hope was that officials would make use of his work to introduce more advanced technology to backward regions, and to disseminate various items of labour-saving or helpful equipment to areas where they were unknown. This motif occurs again and again in the opening sections of the *Nongshu*, for instance in the chapter on harrowing, *balao*, and the many different types of harrow that existed for stirring the mud in rice-paddies, smoothing seed into dry-land furrows, creating a soil mulch, and so on. 'I am including all of them here', says Wang, 'so that northern and southern knowledge can be exchanged, and people can use whatever is most suitable.'[45] And indeed the various types of harrow are all meticulously illustrated and described in the corresponding *Nongqi tupu* section.[46] Unfortunately, Wang's ambition of north-south technical exchange of implements such as harrows, seed-drills or weeding-rakes was never realized, indeed it is possible that it was never attempted by even the most optimistic local magistrate. Wang neglected to consider that these individual tools were embedded in technical systems: a seed-drill, however efficient in the dry soils of the north, simply did not fit with the wet soils and tillage techniques of the south. Perhaps Xu Guangqi (see below) was not entirely wrong to say that Wang Zhen was a better poet than he was a farmer.[47]

In the case of the cotton-processing implements, the gin, bow, spinning wheel and multi-bobbin reels, which Wang Zhen describes for the first time in his 'Illustrated register' (Figure 6.1), his accounts seem to have passed muster even with the exigent Xu, who adds only a few notes explaining minor changes in structure or operation.[48] Here Wang's hope for diffusion was realized: these

43 *Wang Boshan nongshu xu* (Preface to Wang Boshan's Agricultural treatise), *WZNS*: 445.

44 For example the "very complicated" rotating mechanism of the pallet-chain water-pump, *WZNS*: 326.

45 *WZNS*: 27.

46 *WZNS*: 205–207.

47 *NZQS*: 123.

48 *WZNS*: 415–420; *NZQS*: 275–279, and see Shi Shenghan's comments, *NZQS*: 989–990.

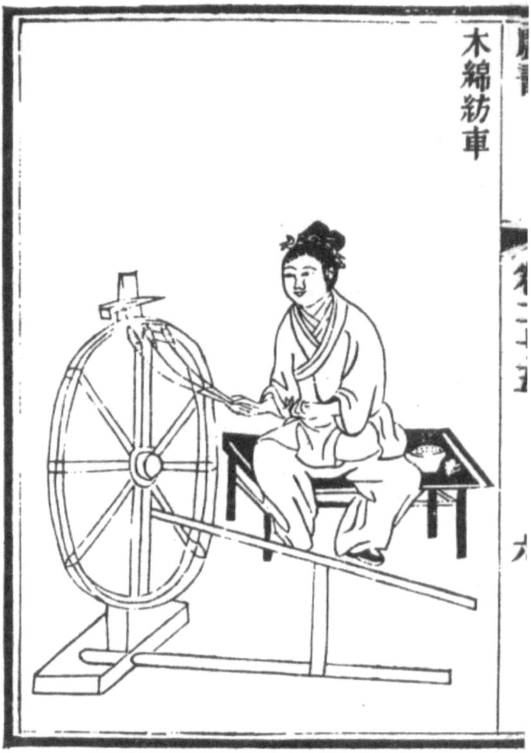

FIGURE 6.1 Woman spinning cotton, 1530 edition of Wang Zhen's *Nongshu*. This treadle-operated machine fills not one but three bobbins at a time.

implements were gradually adopted throughout China, although we have no way of pinpointing exactly what contribution the 'Illustrated register', or official instruction, may have made to this process of innovation, local officials certainly used the work as a reference for providing instruction in processing techniques, or for setting up teaching workshops.[49] But merchants and migrants must also have played a role in disseminating this technical information; Wang Zhen himself mentions that merchants had played a role in the early dissemination of cotton cultivation.[50]

My final examples of innovative official approaches to agronomy in response to crisis come from the distinguished statesman and polymath Xu Guangqi

49 See Kuhn (1988) on state programmes for introducing or improving sericultural techniques during the Song, and Will (1994) on late Ming and early Qing official schools for teaching textile techniques.

50 *WZNS*: 414.

(1562–1633).[51] From the mid-fifteenth to the late sixteenth centuries, the Ming dynasty (1368–1644) flourished: farming prospered and farmers were steadily integrated into national and international networks of commerce; the population increased rapidly, cities grew, markets expanded and print culture exploded. But by the early 1600s pressure on resources and climatic shifts brought hardship and rebellions, weakening the state and exposing its frontiers to invasion. Xu Guangqi, like many of his contemporaries, devoted his life to seeking solutions to the desperate crisis of the late Ming. 'His exhaustive researches on natural and social phenomena were all directed towards *shiyong* [the practical application of knowledge in pursuit of social order] – and what preoccupied him the most were agricultural matters', noted the editor of Xu's monumental *Nongzheng quanshu* (Complete treatise on agricultural administration), published posthumously in 1639.[52] The choice of the term *zheng*, administration, in Xu's title is deliberate. The work addresses three levels of official engagement: national policy, regional organization of infrastructure and local campaigns to improve technical skills. At each level, wrote Xu, 'you will only succeed if you pay attention to the details'.[53] No previous agronomist collected and deployed such a rich and varied range of empirical data as Xu.

Xu began his apprenticeship in farming techniques as a boy when his father, a failed scholar, set up a market garden in Shanghai. As his official career progressed, Xu travelled extensively through China, and he always made time for intensive discussions with local farmers, carefully recorded. As well as the family plot in Shanghai, Xu later purchased land outside Tianjin in north China; he used both farms for comparative experiments with crops and fertilizers.[54]

One issue to which Xu Guangqi devoted considerable effort was the improvement of cotton production, by then China's most important crop after rice. As the long chapter on cotton in the *Nongzheng quanshu* demonstrates, in the three centuries since the *Nongsang jiyao* enormous technical progress had been made in cultivation methods, especially in fertilization (cotton needs lots of nutrients) and in adapting the crop to a wide range of soils; however, cultivation standards were very uneven. In his late forties Xu published a short monograph, 'Comments on cotton', *Jibei shu*, to 'explain the many complications of growing cotton. Fearing that they had not reached all farming households, I had the work printed so as to circulate the information'.[55] The *Jibei shu* was incorporated into the much longer

51 Jami *et al.* (2001).

52 Chen Zilong's preface, *Fanli*, in *NZQS*: 4–5; on Xu and *shiyong*, see Bray and Métailié (2001).

53 Quoted in Li Changnian (1983: 5). Many of Xu's close associates were similarly concerned with the integration of technical improvements into administrative reforms, for instance in the field of water control (see *NZQS*: 337–382).

54 Bray and Métailié (2001).

55 *NZQS*: 975.

chapter on cotton in the *Nongzheng quanshu* (Xu continued to add to his materials till his death), but Xu notes the pedagogical limitations of the written form:

> Some of this information will still not have reached the illiterate, so now I am including four maxims, *yan*. If those who can read will simply teach them these sayings, even women and children will be sure to grasp [the principles] completely: '(1) Select unmixed seed-kernels; (2) sow the seed early; (3) roots deep, stems short; (4) plants well spaced and well banked up'.[56]

In his writings on cotton, as on other agronomic topics, Xu goes beyond recording specific technical methods to seek general principles. He provides detailed discussions of each step in the cultivation process, engaging in close critical dialogue with local works written by landowners as well as official *nongshu*. Xu remarks that while it is tempting to criticize failings in the techniques proposed by earlier authors such as Meng Qi, the compiler of the *Nongsang jiyao*, improvements were continuous and he realizes that his own advice may soon be outdated.[57] Each passage Xu quotes is situated in time and space, its techniques tested against evidence from other authors and against the data that Xu himself has collected from around the country, thus offering a spectrum of practices corresponding to different environments. For each procedure (fertilizing, hoeing, and so on) Xu then provides a set of general considerations and instructions that would apply under any conditions (unlike landowner *nongshu*, which pick out only those which apply locally).[58]

Xu's researches also aimed to reform the *technology*, the socio-economic organization, in which farming techniques were embedded. His proposals concerning cotton production are a good example. Cotton was grown almost everywhere in China, whatever the soils or climate. Northern cottons were the finest, but because cotton fibres snap easily in dry air, spinning and weaving were still largely confined to the humid south and the centre of the weaving industry was Xu's native district, Songjiang. Though cotton was grown in Songjiang, the methods, said Xu, were derided as primitive by farmers elsewhere.[59] Cotton merchants made fortunes from exploiting this regional division of labour, buying raw cotton cheaply in the north, shipping it south to the lower Yangzi workshops, and selling the cotton cloth all over China and beyond. Xu was convinced that

56 *NZQS*: 975. As well as maxims, official agronomists often used simple poems to popularize technical knowledge. Wang Zhen ended each entry in his *Illustrated register* with a short ditty either of his own composition, or by an earlier official like the famous Song statesman Wang Anshi.

57 *NZQS*: 963.

58 For further examples of Xu's innovations in agronomy, see Bray and Métailié (2001).

59 *NZQS*: 964.

farmers across China would benefit if they could both grow good-quality cotton and process it locally; officials should help local farmers match cultivation methods to local conditions and gradually improve the quality of their products.[60] At the beginning of the seventeenth century the peasants of Suning in Hebei had discovered that cotton could be spun and woven during the northern summer if it was done in underground cellars where humidity could be conserved. Xu describes these cellars in some detail, noting that the quantity, quality and price of Suning cloth had risen rapidly in just a couple of decades, reducing the region's dependence on imports. The government, Xu said, should heed these price and supply differentials and devise policies to reduce regional inequalities; in each district it should be possible to manage gradual but steady improvement, culminating in a high level of local self-sufficiency in cloth.[61]

Concluding remarks

Unlike other sciences of state in late imperial China, for instance astronomy and calendrical science, agronomy, *nong*, was a domain of collective, complementary knowledge production that spanned the political spectrum rather than being restricted to specialist imperial bureaux or to members of the educated elite.[62] "Old farmers" or "peasants" were acknowledged by authors of agricultural treatises as primary sources for empirical observations of nature, and as experts in material skills and techniques. Serving officials were expected both to use and to contribute to the agronomic corpus, collecting and processing local expertise for broader dissemination, or introducing suitable methods gleaned from the corpus. Officials were highly educated men, experts in techniques for encoding knowledge in textual or graphic forms, and they had unquestioned authority to elicit technical information from farmers and craftsmen. The authors of the *Nongsang jiyao* documented farmers' techniques of cotton cultivation in order to spread knowledge of a valuable new crop across the empire. Wang Zhen innovated at another level: his *Illustrated register* invited fellow officials to attend for the first time to the construction of agricultural implements and processing equipment, which he encoded in such a way that his educated readers should be able, in partnership with skilled carpenters and other craftsmen, to reproduce these artefacts for adoption in their own jurisdiction.

60 *NZQS*: 970.

61 *NZQS*: 970–971. Late imperial officials often recommended taking advantage of cross-regional comparative advantage (Will, 1994). For more on the discrepancies between official and private attitudes towards *li*, benefit or profit as a factor in decision-making, see Chapter 7.

62 On the varying forms and levels of state control over astronomical and military expertise through late imperial times, see, for example, Waley-Cohen (1999).

Although science (like technology) is a concept that has no direct equivalent in imperial culture, I have argued that agronomic authors strove to identify and explain general principles whereby human action effected material transformations in varying geographical or environmental contexts. Landowner and official authors of *nongshu* made contrasting claims for the validity of the knowledge they produced. Whereas landowners, as we shall see in Chapter 7, strove to provide detailed, precisely situated local knowledge, the goal of official agronomists was to produce mobile forms of authoritative natural knowledge that could be applied anywhere in the empire. They were aided in this quest by the very nature of official experience: officials were regularly transferred from one province to another, and their training and duties required them to become familiar with a broad spectrum of historical documents, local records and memorials on policy, thus encouraging a comparative (if not always critical) approach.

Officials wrote their works primarily for other officials, intending that they should select the advice appropriate to local conditions to instruct local farmers. These farmers were thus at once a source of information and the practising experts who tested or contested the validity of the scientific principles and technical knowledge contained in the *nongshu*. Authors such as Xu Guangqi incorporated objections or failures into a treatment of agronomy that was specifically self-critical and provisional.

Official agronomy tended to occlude the role of literate, landowning farmers, emphasizing collaboration between officials and peasants as the foundation of the social and cosmological order. In practice, however, landowner-farmers were among the (most critical) readers of official *nongshu*, purchasing them from commercial publishing houses and drawing on them selectively and critically in their farming practice and in their production of new treatises. Literate landowner farmers were also important interlocutors of local officials in campaigns to "promote farming". Indispensable as partners in the organization of infrastructural works, they also often took on the role of "instructor farmers" entrusted with the task of disseminating improved techniques among the local population. This brings me to *technology*, the web of cultural and social values and relationships within which science is validated, and to the power of agronomic science as a tool of governance.

In a scathing indictment of modern farming models and their universalist pretensions, James C. Scott offers the interesting suggestion that the urge of modern planners to generate simple, legible landscapes (such as the vast corn fields of Iowa) stems from 'a certain *aesthetic*, what one might call a visual codification of modern rural production and community life'.[63] States need this categoric conformity to govern, says Scott, but oversimplication diminishes the effectiveness

63 Scott (1998: 253, emphasis added).

of government and, by failing to incorporate *metis* (flexible, locally rooted practical knowledge) into its science, it deprives the system of resilience and renders populations vulnerable.

The "good science" of imperial Chinese agronomy reflected an equally potent aesthetic, crystallized in an iconic landscape of tiny fields yielding rich and varied produce thanks to the skilled labours of myriad peasant farmers (Figure 6.2). This aesthetic, like imperial agronomic knowledge, was a joint production of small farmers, landowners and officials, dating back at least to the twelfth century.[64] The Chinese state played a central role in codifying and disseminating the technical knowledge, and in financing and organizing the material infrastructure, needed to implant these ideal landscapes in new territories. But the knowledge the state deployed, like the concrete contours of the landscape, also expressed the knowledge and skills of local farmers, whether peasants or landowners, and broadly

FIGURE 6.2 Fertilizing the seedlings, 1742 imperial edition of *Gengzhi tu*.

64 See Chapter 2.

accommodated their goals. In terms of how technical practices were embedded into a *technological system*, landowners were concerned with *li* in the sense of profit, whereas officials were expected to promote the welfare, *li*, of the local peasantry. Yet these apparently divergent goals of landlord and official science were largely compatible within the context of the imperial rice-based farming system, where landlords did not need to evict tenants and farm vast estates in order to extract the rents they needed, and experienced peasants were known to farm as well as any landlord.[65] Although the state might intervene to control what it regarded as landlord excesses,[66] there was no fundamental contradiction between the interests of landlord farming and state fiscality or approved social order. Thus technical knowledge could circulate freely, as relatively hard facts, between peasants, landowners and officials, and the compatibility of technological codes was certainly a significant factor in the resilience of this iconic landscape, even as the economy it supported developed in complexity and productivity.

I am certainly not the first person to point to the importance of state interventions in farming practice as a tool of governmentality in imperial China. But by recasting the analysis in such a way as to highlight the passages between science, technique and technology, and by relating the making of politically powerful forms of natural knowledge explicitly to the production of the material worlds in which they were embedded, I hope I have demonstrated how this approach might contribute to enriching our understanding of science and rule not only in imperial China, but in pre-modern societies more generally.

65 I discuss the absence of scale economies in skill-oriented wet-rice technology in Chapter 2.

66 For example, attempts to reclaim excessive amounts of farmland from lakes, threatening systems of irrigation and flood control (Shiba, 1998).

7

A GENTLEMANLY OCCUPATION

The domestication of farming knowledge

According to Confucian principles, those who worked with the mind, *shi*, should govern those who worked with their bodies. The governed fell into three ranked occupational categories: farmers, *nong*, artisans, *gong*, and merchants, *shang*. Agriculture was defined as the fundamental occupation, *benye*; crafts and trade were considered at best secondary, at worst pernicious, since they drained labour from the fundamental occupation and encouraged extravagance. To farm was to fulfil one's obligations as a dutiful subject and a person of moral worth. Just as weaving epitomized female virtue for women of any class, so farming was work that men of high as well as humble status could undertake with pride. Conversely, since the labours of farming produced men of moral worth, the path to an official career was open (at least in principle) to the sons of peasant as well as gentry families.[1] 'Working the land', observes Michel Cartier, 'was the only economic activity that did not "debase", the activity that an educated man could and should take up whenever he was not studying or working in the service of the state'.[2] Nor did gentlemen necessarily restrict themselves to the mental labour of managing their farm: some took pride in splashing through the muddy fields with a plough.

1 The moral worthiness attributed to farming was such that throughout the imperial period the examinations were open to candidates from peasant families, although the sons of merchants were debarred until the late Ming and those of artisans until well into the Qing. Although the costs of a classical education were far beyond the means of most farming families, village schools and generous local patrons set many boys of humble background on the path to success as officials or scholars. Whether this translated into significant social mobility in late imperial China has been much debated; see Langlois (2001) for a summary.

2 Cartier (1984: 304).

Literate elites are always convinced that turning the world around them into text produces a superior form of knowledge, and not surprisingly gentleman farmers in late imperial China penned a great many agricultural treatises, *nongshu*. The survival rates of these private *nongshu* were low compared to those of official *nongshu*, for reasons discussed below, but a sufficient number have survived for us to draw some interesting comparisons between the types of knowledge characteristically pursued in each genre, and between two distinctive yet complementary forms of masculine responsibility expressed in the authorial personae of their writers.

To recapitulate the argument of Chapter 6, official *nongshu* were principally concerned with developing and disseminating agricultural knowledge as a tool of government. They were written primarily for the use of other officials as "shepherds of the people", *mumin*, and they tended to formulate technical knowledge in forms that could be transported (and hopefully applied) across the reaches of the empire. These works were promoted, printed and distributed by the government or by local officials, as part of their efforts to promote agriculture, *quannong*, to benefit the people, *limin*, and to maintain the social and moral-cosmological order of the state. Many of their contents, however, entered a much broader social repertory of knowledge and imagery, whether as commercial editions, excerpts in household handbooks and privately authored farming treatises, or images adopted for decoration or for encyclopedia illustrations.

What I call private *nongshu* were works produced by working farmer-landowners. Here the main concerns and technical strategies reflected the challenges of keeping the family farm or estate going. The authors were concerned to generate detailed, specific knowledge adapted to local conditions, and they wrote for their kinsmen, their descendants and other landlord-farmers in the region. Often these works incorporated materials from official *nongshu*, but usually such materials were substantially edited or rearranged for inclusion. Many of the private works were never printed, though often they circulated quite widely as manuscript copies.

Official and private works drew on a common pool of agronomic materials, and new knowledge generated in one category was often taken up in the other, yet the goals, conceptualization and presentation of knowledge differ significantly between the two genres. To illustrate some key characteristics of private *nongshu*, I present three roughly contemporary works: the *Shenshi nongshu* (Agricultural treatise of Master Shen, hereafter *Shen's treatise*) of c. 1640; the *Bu nongshu* (Supplemented treatise on agriculture [of Master Shen], hereafter *Supplement*), by Zhang Lüxiang, completed in 1658; and the *Nongsang jing* (Canon of farming and sericulture, hereafter *Canon*) by Pu Songling, whose preface is dated 1705. All three *nongshu* were written within a few decades; the first two are by Jiangnan authors, while Pu Songling, a native of Shandong, was writing in the tradition of northern farming.

One reason for selecting these works is that they have survived more or less intact and are available in modern, critical editions;[3] another is that they were completed not long after the agronomic writings, both public and private, of the eminent late Ming statesman Xu Guangqi. Not only was Xu the author of the monumental and innovative *Nongzheng quanshu* (Complete treatise on agricultural administration, hereafter *Complete treatise*), discussed in Chapter 6, but he also was a practising landowner-farmer in his own right. Nevertheless, the obligations and epistemologies of the statesman, the *mumin* or shepherd of the people, always predominated in Xu's agricultural writing, even when he was writing about local practice in the kind of detail that was more typical of landlord farmers. Xu's style of agronomic writing thus highlights some of the key differences between public and private styles of agronomy, their methods and goals, and the social role or persona of their authors.

Private works

The main category of works that I consider under this heading are farming treatises written by working landlords. Wang Yuhu's bibliography of agricultural works throughout Chinese history counts 105 works on agriculture that were attributed to the Song dynasty, 26 to the Yuan, and 310 altogether to the Ming and Qing. The majority of these were short works by local, private authors, most of which we know only by the inclusion of their titles in old bibliographies, or by one or two quotations in surviving works.[4] Some of these works were printed on or soon after their completion, others not. Lucille Chia, in her comprehensive study of the printing industry of Jianyang in the Song, Yuan and Ming, finds records of only one work published by a commercial house that falls under the *Siku quanshu* heading of agriculturalist, and that was published in the Song. Even though Jianyang was just one book-publishing centre of several in late imperial times, and even though the Jianyang publishers were not active during the Qing, this still suggests a rather small market for printed works on farming, since Chia finds a total of 278 titles published in Jianyang on medicine, 244 during the Ming alone.[5] On the other hand, household encyclopedias all contained sections on farming – often rather generic in the information they contained. I would venture to suggest that the later Ming and Qing households that were avid purchasers of printed works on medicine, philosophy or examination techniques were not uninterested in texts on farming, but preferred those which were tailored to local conditions and which

3 Chen and Wang (1983); Li Changnian (1982).

4 Wang Yuhu (1979).

5 Chia (2002: 312).

may very well have been circulating not in printed form but in manuscript, no doubt heavily annotated by each owner.

Shen's treatise and the *Supplement* were written soon after Xu Guangqi's collection of materials for what was to become the *Complete treatise*, and the authors were from prefectures very near to Xu's Shanghai home: Master Shen lived in the district of Gui'an in Huzhou prefecture, and Zhang Lüxiang in the contiguous district of Tongxiang. About Master Shen we know almost nothing except that he probably wrote his work just before 1640; we can tell from his text that he was a working landlord who paid the land tax, which in the late Ming may well have meant that he had no degree.[6] Zhang Lüxiang (1611–1674) came from a gentry family that had fallen on hard times. He was a promising scholar, passing the *xiucai* degree at 15, but he never passed the *juren* exam. Instead he turned to teaching, and later resumed his studies under the Confucian reformist philosopher Liu Zongzhou. Like Liu, Zhang was an uncompromising Ming loyalist, and in the years following the fall of the Ming he lived quietly on his lands, teaching in the village school, farming and corresponding with other loyalist scholars from the region.[7] In 1647 he transcribed *Shen's treatise*, and by 1658 he had completed his own additions, the *Supplement*. Here is the rationale that he gives in his afterword to *Shen's treatise*:

> This book probably comes from Master Shen of Lianchuan, and was completed at the end of the Chongzhen era [1628–1644]. The conditions are not very different from those in my own district. Master Shen's descriptions of the various tasks of cereal cultivation, sericulture and animal husbandry are all methodical and precise, beyond what any experienced farmer or raiser of silkworms has yet produced. The monthly ordinances, *yueling*, at the beginning of the work go right to the heart of the principles of proper timing. The following sections laying out different tasks are detailed and full, for he has taken great care to be comprehensive.
>
> I myself have spent a number of years studying farming, enquiring carefully into successful and deficient methods, and making myself fully acquainted with its principles. But as a youth I did not practice ploughing

6 Chen and Wang (1983: 1) base this dating of Shen's work on the prices he gives for grain, silk and other commodities. They deduce his educational status from his complaints about having to pay the land tax (Chen and Wang, 1983: 4). An alternative interpretation is that Shen did have a degree but was complaining about tax payments for which he was liable, albeit at a reduced rate.

7 Chen and Wang (1983: 3–6). After the fall of the Ming, Liu Zongzhou (1578–1645) refused office under the new Qing dynasty and starved himself to death; Zhang mourned his loss bitterly (Chen and Wang, 1983: 3). It is perhaps not unrelated to Zhang's decision to write a *nongshu* that Liu was one of a 'group of refined scholars [of the late Ming] . . . who turned their attention to nature and material inventiveness' as part of a programme of investigations of how *qi* could be mobilized to reform society (Schäfer, 2011a: 16).

> myself, so my bones and sinews lacked strength and I hired labourers to work my land for me. But their work was very rough, and left to themselves they did not seek for finer investigations – it was as if I had stony land! Thereupon I wrote out these chapters, and clarified their meaning together with my family members – this is far sounder than relying on verbal explanations alone. The *Book of Poetry* says: 'Even though we lack old, wise men we still have rules and statutes.'[8]

Shen's treatise occupies about 90 pages in the modern edition. It consists of a monthly calendar of general tasks, followed by a long section on crop cultivation, a section on sericulture, and a brief section on "daily necessities", which contains several recipes for preserves and such. It is possible that it was Zhang who arranged the text in this fashion while transcribing it. The *Supplement* is about 50 pages in the modern edition; the first section consists of various entries on specific practical tasks, while the second contains a variety of general reflections on farming.

In Zhang's preface to the *Supplement* he says that after he had copied out *Shen's treatise* another scholar from the Tongxiang district, Xu Zijing, remarking that *Shen's treatise* was incomplete in some respects, asked Zhang why he did not write a supplement.[9] Xu too was a Ming loyalist and he owned 400 or 500 *mu* of land, the rents on which provided most of his income – he was thus the kind of social superior that one would ask to write a preface or to edit one's work. Zhang had indeed noted that there were differences between Shen's district and his own, which meant that some of the details of Shen's work on Gui'an did not apply in Tongxiang. So Zhang jotted down the main lines based on his discussions with experienced farmers, and Master Xu, according to Zhang, edited them. Yet, says Zhang, 'in farming there are the roots and the branches, the fundamentals and the secondary matters. Master Shen has perfectly covered all the fundamentals; what I say simply deals with some secondary matters'.[10]

The modern edition by Chen Hengli and Wang Dacan is based on eight surviving woodblock editions, some of *Shen's treatise* alone, others of the *Supplement* alone, and some combining both texts. It seems that Zhang Lüxiang's various literary and scholarly works were not actually printed during his lifetime. Soon after his death, his disciple Yao Sixia collected and transcribed them under various headings, including Zhang's hand-written version of the *Supplement* (but apparently omitting *Shen's treatise*) in a collection that was printed under the title *Yangyuan xiansheng quanji* (Complete works of the Master of the Willow Garden). The circulation of this edition was very limited, but in 1704 Zhu Renzhai, a scholar from nearby

8 *BNS*: 97–98.
9 *BNS*: 99–100.
10 *BNS*: 100.

Haining, printed a version entitled *Yangyuan quanji*. This was re-issued in mid-Qianlong (c. 1760) by a Suzhou scholar Zhu Kun, who also included *Shen's treatise*. When the imperial collection *Siku quanshu* was compiled and issued in 1773, it included both a *Zhangyuan quanshu* under the heading of "various philosophers", and *Shen's treatise* under "agricultural masters" – the latter being based on a copy in the family collection of a Hanlin academician called Cheng who was engaged in the imperial compilation. Subsequent editions seem all to have included both *Shen's treatise* and the *Supplement*. At the same time, the combined work seems to have enjoyed a fair degree of popularity among practising farmers, not only locally but in surrounding regions, becoming quite popular further north in Jiangsu, and upstream in Anhui and Jiangxi.[11]

Pu Songling (1614–1715), from Zichuan in Shandong, is of course best known as a writer of fiction, famous in particular for *Liaozhai zhiyi* (Strange Tales from a Chinese Studio).[12] Pu wrote relatively little on agriculture, or at least little that has survived: the *Nongsang jing* (Canon) itself, and the *Nongsang jing cangao* (Chaff from *Nongsang jing*, hereafter Chaff), both of which are included in Li Changnian's modern edition.[13] As far as we know the *Canon*, prefaced 1705, was never printed; however, as soon as it was completed it was apparently passed around in the countryside, and soon it was being copied – so there are still several hand-written copies in existence, with some minor differences but essentially the same. This, says Li, is because it reflected local farming conditions and so it appealed to people. Li drew upon seven surviving hand-written copies for his edition, together with a version included in Lu Dahuang's 1962 *Pu Songling ji* (Collected works of Pu Songling).[14] We might note that the *Canon* is arranged by month, in the form of a *yueling* ("monthly ordinances", see below), and yet the last three months of the year do not figure in the main text although there are several entries under months 10, 11 and 12 in *Chaff*. Presumably Pu Songling never got around to finishing the *Canon* itself.

Shen's treatise and the *Supplement* seem to consist primarily of original materials – at any rate, if there are quotations from earlier works they are unattributed. This is a very different tactic from the typical official *nongshu*, where a more classical display of authority and expertise is the norm: as many works are quoted and acknowledged as possible. Pu Songling has one earlier authority whom he considers an excellent source for knowledge on local conditions, and whom he quotes freely. At the same time, he is concerned to make sure that the wisdom he transmits is truly suitable for local conditions:

11 Chen and Wang (1983: 2, 1).

12 Pu Songling (2006).

13 Li Changnian (1982).

14 Li Changnian (1982: 3, 5).

> The essential tasks for householders boil down to farming outside, and sericulture inside. Long ago Master Han wrote his 'Farming Instructions', a work so clearly expressed that even silken-breeched dandies and book-festooned scholars could understand the farming matters [it describes]. I read the work and thought it excellent. However some of what it says is incomplete, while other parts apply in one place but cannot be applied in another; I have therefore added to and cut [the original text]. I have also taken excerpts from a wide range of ancient and modern treatises on sericulture and collated them into a single book, that goes after [the book on farming]. Although this work cannot transform the whole world, *tianxia*, many people may like to hand it down to their sons and grandsons.[15]

Perhaps because the work was in perpetual progress, Pu Songling does not provide attributions for all the passages he quotes from earlier works. It seems that a work by one Master Han, namely *Hanshi zhishuo* (Plain talking by Master Han), was indeed one of his principle sources. This is one of a series of works devoted to northern dry-land farming that dates back to the Song, Jin or early Yuan. Among these works we can also count the *Zhongyi biyong* (Practices necessary for growing crops) of the Song;[16] the Jin or early Yuan work *Zhongshi zhishuo* (Plain talking on agriculture);[17] and three short works from the early Qing, *Zhiben tikang* (Selected guidelines for understanding the fundamental occupation), *Nongyan zhushi* (Farming sayings set forth) and *Mashou nongyan* (Farming sayings of Horsehead District).[18]

All these works were specialized applications of farming knowledge to the harsh conditions and dry-land techniques of the north, and the sixth-century *Qimin yaoshu* (see Chapter 2) was a primary source for all of them. One key reason for the enduring reputation and long-term survival of *Qimin yaoshu* was the care that Jia Sixie, the author, took to translate the technical advice he provided into mobile knowledge that could be adapted to local circumstance, systematically describing, for example, which combination of tillage methods best suited which kind of soil,

15 *NSJ*: 3.

16 Compiled by Wu Cuan, probably during the Song, with a supplement, *Zhongyi biyong buyi*, composed by Zhang Fu, probably in the late thirteenth century around the time of the Jin-Yuan transition. These paired works fell from sight for centuries, but were identified in the *Yongle dadian*, the Ming imperial encyclopaedia of 1408, by the agricultural historian Hu Daojing, who published a critical edition in 1963. Wang Yuhu (1979: 105–106) characterizes them as compilations in the *biji* ("jottings") style, valuable because they record popular practices, including beliefs concerning auspicious days for planting different crops (Hu, 1963: 63).

17 This may very well have been the *Hanshi zhishuo* under a slightly different title (Wang, 1979: 106–107).

18 The first work, by Yang Cui, was published in 1747; the second, by Yang Xiuyuan, sometime in the later Qing, and the third, by Qi Junzao, in 1836. These three rare works were anthologized by Wang Yuhu (1957).

or how sowing rates should be adapted to soil, season and climate.[19] Although Jia claimed in his introduction that his book was written for his own children, it offered knowledge in forms that seem to have been equally attractive to the successive governments that ordered copies to be distributed to serving officials, and to private landowners. While most would probably have just have marked up the margins of their copies with their own observations, qualifications and criticisms, others – including the authors of the works just listed – used *Qimin yaoshu* as a structuring device, quoting passages and then adding criticism, qualifications and additional refinements based on local conditions and experience.

This kind of local knowledge, however, was less likely to travel. It is not surprising that many of the private *nongshu* have been lost, while others survived either by accident, or because their authors were sufficiently respected as experts in other fields to have their works collected or anthologized. While a few, such as *Qimin yaoshu* and Chen Fu's *Nongshu* were of such path-breaking quality and originality that they achieved canonical status, *Plain talking by Master Han* and *Plain talking on agriculture* are by no means unusual in having been preserved only in references and quotations. Wang Yuhu suggests that interest for these works among educated readers was displaced by the production by the Yuan Agricultural Ministry in 1273 of the official compendium *Nongsang jiyao*. Passages attributed to both works occur frequently as quotations in both the *Nongsang jiyao* and in Wang Zhen's *Nongshu*, thus assuring their partial transmission through to modern times.[20] They were also taken up again by Pu Songling, and various passages were incorporated into a Ming compendium entitled *Zhongshu shu* (Manual of arboriculture) by Yu Zongben.[21] However, as Wang Yuhu notes, they were not cited in any southern *nongshu* of the time or later, because they describe farming practices from North China.[22]

If we consider the surviving private *nongshu* from late imperial times as a corpus, we see that the main *raison d'être* of these private works was to provide specific, local knowledge. In other words, quoting authors who wrote about other environments would be beside the point. Pu Songling refers to knowledge that is excellent but 'applies in one place and not in another'. Zhang Lüxiang remarks

19 Jia Sixie was able to formulate his knowledge in this way precisely because he was an educated and travelled official. He was thus familiar with farming in several regions, as his references to his own experiences and experiments make clear, and he quotes from over 150 works (Bray, 2001a).

20 Chen Fu's *Nongshu* also doubtless owed much of its renown to the fact that Wang Zhen included it in his own *Nongshu*, in this case quoting it in its entirety, albeit in sections distributed between the relevant chapters of his own work. Chen Fu's treatise, like the *Zhongyi biyong*, was included in the *Yongle dadian*, and it also featured in several Qing collectanea, but according to Wang Yuhu it seldom circulated as a separate book (Wang Yuhu, 1979: 86). Nevertheless, through its inclusion in Wang Zhen's *Nongshu*, Xu Guangqi's *Nongzheng quanshu* and official agronomic compendia like the *Shoushi tongkao* (Compendium of works and days) of 1742, the text itself circulated widely.

21 Wang (1979: 107); Menzies (1996: 545).

22 Wang (1979: 106–107).

that although Gui'an and Tongxiang are only a few miles away, in Tongxiang the land is low-lying and damp, and is not suitable for working with animal-drawn ploughs and harrows using the techniques described by Master Shen.[23] There were also differences in the ideal size and organization of the working farm and of tenancies and labour hire, as well as differences in farming techniques and crop choices. The paired works by Shen and Zhang are embedded in a dense web of local knowledge: other landlords' experience, peasant sayings, soil conditions, water supply, rents and wages, prices and profits, are compared and analyzed in detail.

In the late Ming and Qing, the most important and productive improvements in farming technique were often very modest yet crucial changes, for instance the reduction of the size of irrigation units allowing for more precise water control, improvements in the type and timing of fertilizer applications, or the breeding of crop varieties particularly well suited to local conditions.[24] These kinds of improvements represented exactly the kind of local knowledge that authors such as Shen and Zhang were interested in. Their remarks on fertilization, particularly on a new technique called "carrying on the power", *jieli*, that more and more Jiangnan farmers were trying out during the late Ming, show that their writings were just one element in a more diffuse local debate about technical improvement.

Lime and ground shells for soil improvement were widely purchased by rice farmers of the time, as was human manure or nightsoil collected from the towns; Master Shen says that farmers in his district had to go as far as Hangzhou, almost 80 kilometres away, to buy nightsoil. Perhaps the most widespread commercial fertilizer was oil-cake, fibrous wheels left from the pressing of soya or other seeds used for cooking oil. A single finely pounded oil-cake would fertilize 1 *mu* of rice seedlings, enough to transplant an area of 25 *mu*; the use of oil-cake instead of manure was recognized not only to increase yields, but to save labour.[25]

One technique that was advanced, if not perfected, during the Ming was the application of supplementary dressings of fertilizer; Jiangnan farmers referred to this as "carrying on the power". In addition to manuring the main field before transplanting and fertilizing the seed-bed with liquid manure, practices already described by Chen Fu in his *Agricultural Treatise* of 1149, in order to "carry on the power" fertilizer was applied again after the rice had flowered. A Jiangnan author wrote some decades before *Shen's treatise* was completed:

> You should not put down manure too early or its strength will not last . . . Only at sowing time must river mud be applied as a base, and although its strength is lasting and dissipates slowly, by midsummer you should apply a

23 *BNS*: 101.

24 Li Bozhong (1998b); Shiba (1998).

25 *BNS*: 35, 56; see also Xue (2005) on bringing nightsoil from towns to villages.

> little potash or oil-cake which also dissipates slowly and is long-lasting. Only at the end of the summer or the beginning of the autumn should you apply nightsoil, by which time it will have double the effect, so that the rice panicles will grow very long.[26]

As this passage indicates, it was easy to make mistakes, whether in the timing, the quantity or the nature of fertilizers. By the end of the Ming, Master Shen was pointing out that amounts and timing of supplementary fertilizers depended on the weather and the state of the crops; poor farmers who could not afford the supplements would suffer from meagre harvests, but farmers who used too much fertilizer indiscriminately would end up with lots of straw and empty husks.[27]

It is interesting that whereas late imperial official works made heavy use of illustrations (see Chapter 8), most private works depended on text alone to convey information. This is no doubt related to the fact that they were often not intended as works to send for printing: they were bundles of notes jotted down by hand and passed around to neighbours, or down to children and grandchildren. Furthermore, the kinds of illustrations typical of the official works were probably not considered specially relevant. If these private works were for local consumption, there was no need to draw equipment since it was familiar to all, while the iconic value of works such as the *Gengzhi tu* (Pictures of tilling and weaving) was also of very minor relevance to the private works, which were strictly pragmatic and contained little about rituals, cosmic order or the ethical obligation to recognize the dignity of peasant labour. Even in the prefaces, the pieties about the importance of agriculture as the foundation of society were reduced to a minimum.

One ordering device that was often used in private works was the *yueling* or "monthly ordinances".[28] The *yueling* sets out the various tasks of each month of the year, often relating them to natural phenomena such as the flowering or fruiting of certain trees, the movements of the stars, or the appearance of insects or birds. *Yueling* are a very ancient form in China, the very earliest referring to hunting, fishing, and rituals as much as farming. Agricultural writers of the early periods frequently used *yueling* or *yueling*-like linking of farming tasks and natural phenomena to help overcome geographic distance: if instructions were given to plant millet when the apricots first flowered, rather than in terms of the month itself, then northern and southern farmers would both plant at the right season. Official writers such as Wang Zhen and Xu Guangqi struggled to produce diagrams of infinitely transposable *yueling* that could be used anywhere if they were adjusted according to the correct principles.[29] But in works such as *Shen's treatise* or the

26 *Wuxing zhangguji* (Selected historical records of Wuxing), quoted in Chen Zugui (1957: 106).

27 *BNS*: 35.

28 See Bray (1984: 52–55).

29 See Chapter 6.

Canon, the first starting with a *yueling*, the second organized throughout on that model, there was no need to link farming tasks to natural phenomena. Because they were conceived as applicable to one locality, tasks could simply be set down as corresponding to a specific month, thus making the *nongshu* very convenient to use.

Finally, let us turn to the economic sensibilities of private *nongshu*. In official *nongshu* the term *li*, meaning benefit or profit, is typically used in the very general concept of *limin*, benefiting the people; in the private works, however, it typically refers to profits.[30] Private *nongshu* were written by working landlords for other working landlords. They were thus concerned with matters that are usually completely absent from the official works, in particular costs and prices, and issues concerning tenants and hired labourers. Master Shen and Zhang Lüxiang were typical in this respect. Shen hired both male and female workers, for farming and for commercial silk production respectively. Zhang farmed about 10 *mu* of paddy and several *mu* of dry land each year, working with his labourers in a straw hat and sandals to transplant the rice and bring in the harvest. The rest of his land he hired out to tenants.[31]

This was typical: as discussed in Chapter 2, in the skill-oriented farming system that prevailed in China's southern rice regions, even large landowners would give most of their land out in small lots to tenant farmers, keeping only a small home farm for themselves in the case that they did not simply live off their rents. 'In my district even a superior farmer can only manage 10 *mu* of land, wet and dry, so if a family has a surplus of land they will let a tenant farm it and take his rent'.[32] But tenants had to be carefully selected, as did farm workers. The ideal tenant should not need supervision, since his reward for working well was a share of higher yields and profits. But it is doubtful that most landlords (or their bailiffs) trusted tenants any more than the tenants trusted their landlords. As for labourers, since they were usually paid by the day Pu Songling's opinion that they were lazy, greedy and had to be watched like a hawk was probably widely shared.[33] Discussions in private *nongshu* cover not only wages, but also how often the workers should be given meat or wine – although resentful labourers were not as dangerous as resentful carpenters, it was still best to provide regular tokens of appreciation.[34]

30 For a recent, nuanced analysis of how *li* was likely to be interpreted by gentry and by officials, with case studies of the complex negotiations between private interests and public good that framed such vital issues as hydraulic policy, see Li (2012).

31 Chen and Wang (1983: 5–6).

32 *BNS*: 148.

33 *NSJ*: 21, 25.

34 Clients hiring carpenters were terrified that they would use sorcery to build bad luck into their house during construction or repairs. To avoid this they were advised to offer polite treatment and regular meals of meat, and not to stint on wages or materials (Ruitenbeek, 1993: 111).

In a good example of the kinds of calculations working landlords had to make, Master Shen compares the wages of male labourers hired for field-work with those of girls or women hired to help raise silkworms and to reel thread or weave cloth, concluding that the women are more likely to earn their keep than the men. His calculations involve what were obviously locally specific ball-park figures for the prices of rice, of warp and weft threads and of finished cloth – along with discussions of how often the labourers should get meat and wine as well as more basic foods.[35] All are discussed in terms of strings of cash and ounces of silver; in other words for Shen, as for other authors of private *nongshu*, the word *li* refers to a profit that can be rather precisely calculated in monetary terms, within a specific market.[36]

Xu Guangqi: astride traditions

Today Xu Guangqi (1562–1633) is best known as a high-ranking statesman who was a Christian convert and one of the first promoters of Western science in China. He is also known as a loyal official who devised a host of policies he hoped might save the Ming dynasty. Xu has attracted the attention of numerous historians of science for his collaborations with Matteo Ricci, including a famous translation of Euclid, and for his role in calendrical reform.[37] Fewer have recognized the importance in Xu's thought of agriculture. If Xu devoted enormous amounts of time and energy to researching and writing on agriculture, it is because he considered this domain of knowledge to be the core of *shixue*, the "practical studies" which alone could regenerate the Chinese polity and restore social order. Xu himself believed that his greatest contribution to knowledge was his monumental work *Nongzheng quanshu* (Complete treatise on agricultural administration). The work was published posthumously, in 1639, by a group of young officials who shared Xu's belief that promoting agriculture, *quannong*, lay at the core of successful statecraft

35 *BNS*: 84.

36 In the highly commercialized market networks of late imperial Jiangnan we are not surprised to see landowner authors of *nongshu* offering such calculations. This does not signify, however, that these landowners operated their enterprises according to the rationalities of modern business (e.g. Huang, 1985; Hamilton and Chang, 2003). Records of similar calculations date back at least to the *Qimin yaoshu*. Jia Sixie too was concerned with how a landowner could make a decent and sustainable living from farming. His work is full of prices and calculations. If you live near a town you can grow vegetables for profit, but if you live in the depths of the countryside transport will be too expensive. A plantation of 10 *mu* of elms will fetch nothing for the first few years, but then each tree will bring so much a year in firewood, and at last a larger sum for the timber. Calculate carefully how much grain you should sell after the harvest when prices are low, advises Jia, and how much you should store till prices go up (Bray, 2001a).

37 Jami *et al.* (2001).

and was just as essential as military improvements for saving the crumbling dynasty.[38]

It is because Xu Guangqi was both a passionate farmer and a dedicated official that his *Complete treatise* is such an extraordinary and original work. Xu came from Shanghai. His father, Sicheng (1535–1607), was the only heir of a relatively well-off family of scholars and traders, but by the time Guangqi was born Sicheng had been reduced to poverty and was earning his living as a vegetable gardener; Sicheng's mother and wife worked all year round spinning and weaving to help support the family.[39] Historians presume that Xu Guangqi learned the basics of horticulture from helping his father as a child; he also acquired a healthy respect for the usefulness of cotton, which in the course of the previous century had transformed the economy of the Shanghai region. Although poor, his family scraped together the money to send their only child to school at the age of six. There he immediately declared his Confucian commitment to becoming a servant of the state and working for the benefit of society, 'bring[ing] order to the state and the people'.[40]

Through Xu's youth the family had almost no land that could generate an income, only a house plot inside the city and a small garden plot outside the walls. Their livelihood was precarious, and in the famine year of 1581–82 Xu's wife sold some of her modest dowry to allow him to continue his studies.[41] Xu passed the *jinshi* degree in 1604, after which he served in a sequence of prestigious official posts at the capital. Back in Shanghai he purchased an estate and villa in Longhua, and his letters from the capital to his son Ji are full of advice and instructions about managing the property.[42] Towards the end of Xu's career his Shanghai farmlands provided much of the rice he consumed in the capital, although with the constant threat of natural disasters and raids by pirates, this income could not be relied upon.[43]

In 1613 Xu bought a second estate in Tianjin, where he spent several periods performing experiments with rice cultivation – and increasing his income. He first had to clear the land, and in a long letter written around the time of the purchase

38 The goals, structure, contents and innovative contributions of the *Nongzheng quanshu*, and the processes of its compilation and publication, are discussed in Bray and Métailié (2001).

39 Wang Zhongmin (1981: 6–7); Übelhör (1969: 29–42).

40 Brook (2001: 94).

41 Brook (2001: 95).

42 Wang Zhongmin (1981: 88–89). It seems that at the time Xu took his *xiucai* degree he owned only a few measures of overgrown land that he cleared and planted with willows; the plot generated a greater profit from firewood than it would have done in rent (*NZQS*, *Zhang Pu yuan xu*: 2). Jia Sixie would have approved.

43 According to P.-E. Will, the irregularity in the payment of official stipends may be attributed to the deepening financial crisis and administrative disorder of the late Ming; still worse, by this time official salaries were insufficient to support even an austere official if he was honest (Will, personal communication, 2000).

he describes the environment, its possibilities and problems, including the lack of labour. In one letter he writes that since the climate is so dry, 800 *mu* under wheat will barely yield enough to cover the land tax and give a tiny profit, for each *mu* will yield 5 pecks, *dou*, of which the landlord's share is only 2 pecks, amounting to a 150 or 160 bushels, *shi*, in all. However, once Xu had converted his lands to irrigated rice, the income increased. According to one letter, the 40 per cent of rent amounted altogether to some 300 or 400 bushels of grain a year; another letter mentions sending 100 taels of silver home to Shanghai for his son to invest.[44] Xu also used both his Shanghai garden and his Tianjin estate for field experiments, which he carefully wrote up.

Hu Daojing considers that Xu's agricultural writings fall into three phases:[45]

1 His earliest works were monographs based on his experiments in Shanghai between 1607 and 1610, when he was on leave from his official duties, mourning his father. These treatises on sweet potatoes, turnips, cotton and bamboo synthesized the results of his personal experience and experiments, and the practical information he had collected from other cultivators.[46] The purpose of the research and writing was to introduce new famine crops (sweet potatoes and turnips) to the Shanghai region, and to popularize the best ways of cultivating important local economic crops.
2 Based on the experience of several extended stays on the Tianjin estate (three and a half years altogether between 1613 and 1618, with two brief final visits in 1621), Xu wrote half a dozen works on land management and rice cultivation in North China. Xu was convinced that the frontier regions could feed themselves if rice farming military colonies, *tuntian*, were set up. He purchased his Tianjin estate as an "experimental example", *shefan*, on which to base his policy recommendations.[47]
3 In his later years Xu wrestled with the drafts of the grand treatise that was to become the *Nongzheng quanshu*.

44 The letter discussing the purchase is in Wang Zhongmin (1984: 487). The letter on wheat is quoted by Hu (1980: 126). Wang Zhongmin (1981: 94) cites other letters to produce the figure of 300 or 400 bushels of rent a year.

45 Hu (1980).

46 Xu had a vast appetite for the practical details of farming and for the intricacies of local variation in agricultural practice. He travelled widely, and his son Xu Ji wrote that he was relentless when questioning farmers, writing down their replies in detail and never stopping until the subject was completely covered (cited in Shi Shenghan, 1979: 4). These enquiries were an important source for many of Xu's personal contributions to the *Nongzheng quanshu*, seen in passages that begin in such ways as 'In Songjiang they . . .' or 'The people of Wu . . .'.

47 Li Changnian (1983: 4); for the background to this project, and its failure, see Brook (2005c) and Bray and Métailié (2001).

To these we should add Xu's extensive researches and writings throughout his career on the administrative dimensions of agrarian and fiscal problems.[48] These ranged from policy papers on *tuntian* to Xu's research on locusts. In the heroic period when many historians were prone to seek "proto-science" in pre-modern China, Xu's experiments and researches were often identified as examples of "scientific spirit and method".[49] Although Xu's researches were methodical and rigorous, they owe more to Chinese tradition and less to contact with Western learning than the term "scientific" implies in this context. Xu's methods were closer to evidential scholarship, *kaozheng*, than to "proto-science": starting with an exegesis of ancient texts, he then tested the facts on the one hand, and introduced new elements on the other. As a Hanlin Academician, Xu had privileged access to a huge corpus of historical records and texts, and he put the materials he collected to continual use in generating new knowledge.

A good example is Xu's research on locust control, based on a systematic study of all the available historical and contemporary records of locust swarms. This enabled Xu to locate their origins (in marshy areas), the climatic conditions under which they were most likely to swarm, and their behaviour during a swarm – from which he was able to suggest methods for locating and killing the locusts early in the cycle, thus averting most of the damage. However modern this approach may seem in hindsight and with some editing out of awkward beliefs,[50] it seems unlikely that Xu's research method owed anything to the Western knowledge brought by the Jesuits, whereas the perusing of historical records to discern patterns was a time-honoured method in Chinese scholarship. One of the most significant points about Xu's work on locusts was that it translated into effective action – his methods were widely adopted. Another is its guiding philosophy, which Xu felt applied not only to locusts but to every form of disaster, from a local famine to the collapse of the state: 'prevention [of disastrous conditions] is the best strategy, preparation [against disasters] is [only] second-best, and [disaster] relief should be a last resort'.[51]

One can imagine another dedicated official producing archival research similar to Xu's study of locusts. His research on sweet potatoes and turnips was much more hands-on; it would not have succeeded without Xu's work in his own experimental plot, nor without his intimate knowledge of the Shanghai ecosystem. To this extent it resembles the kinds of personal testing and adaptation to local conditions typical of private *nongshu*. Yet without Xu's broad geographical

48 See Übelhör (1969) and Bray and Métailié (2001).

49 See particularly Li Changnian (1983).

50 Xu shared the popular understanding that locusts turned into frogs in winter and spent the cold months buried in mud at the bottom of ponds. In Britain this belief applied not to locusts but to swallows; it was still in force among naturalists as late as the 1780s, as we learn from Gilbert White's *Natural History of Selborne*.

51 *NZQS*: j.44, quoted in Li Changnian (1983: 8).

experience, acquired as an official, he would not have recognized the potential of certain exotica for local adoption. Xu was acutely aware of the vulnerability of Jiangnan farmers to floods, droughts and consequent crop failures. A central theme of the *Complete treatise* is the need for farmers to diversify, and to plant crops that would save them from starvation in a bad year, but could be put to other uses if the year was good.[52] But often farmers rejected new crops on the grounds that they would fail to thrive outside their region of origin. One driving force behind Xu's crop research was to prove that this principle was mistaken: on the basis of the trials he conducted on his own land, he urged introducing rice from Jiangnan to the northern frontiers, and sweet potatoes from the tropics as well as northern turnips to Jiangnan. In the last two cases at least, he was successful.

There was a common saying that the sweet orange, native to the South, metamorphosed into a thorny citrus when it was planted north of the Yangzi – in other words, man proposes, nature disposes. *Wujing* turnips (*Brassica rapa*, a member of the cabbage family) were a common crop around Beijing, and Xu saw the potential for using the starchy roots as a staple in Jiangnan at times when the rice crop failed. However, Jiangnan people claimed that the *wujing* turned into a *song*, a leafy cabbage (*Brassica chinensis*), when planted south of the Yangzi. The turnip and the cabbage both belong to the same family and resemble each other rather closely, but the turnip normally has a much larger root than the cabbage. Xu pointed out that it was mistaken cultivation techniques that led to the problems with *wujing* in Jiangnan. Northern farmers fertilized their turnips heavily so that the roots would develop. But in the south it was customary to give very little fertilizer to *song*, which were used as green vegetables. Southern farmers who tried to grow turnips followed the same local practice, so not surprisingly their turnips ended up with small roots like *song*. Yet turnips could be grown very successfully south of the Yangzi, as Xu himself proved in his experimental garden, provided the right cultivation techniques were used. Xu used the example of the turnips, and also of sweet potatoes, to prove that plants could adapt to almost any conditions provided the right cultivation techniques were found.[53]

The results of Xu's experiments with crops were incorporated into the relevant sections of the *Complete treatise*, whose structure followed the conventional pattern of official *nongshu*, beginning with a lengthy section entitled *Nongben* (Agriculture is the foundation), followed by field systems, farming methods, water control, crops, textiles and animal husbandry. But it should be noted that the first section, *Nongben*, covered a series of vigorous contemporary debates on policy in which Xu and his colleagues were engaged. These included agricultural colonies, *tuntian*,

52 Bray and Métailié (2001).

53 *NZQS*: 717 and Hu (1980: 120) on the tale of the turnip and the cabbage. On Xu's successful strategy for introducing sweet potatoes, see *NZQS* j.27 and Bray and Métailié (2001).

and the intricacies of provincial-level and local irrigation reform. Also reflecting the urgent needs of the time, the work ends with a very long section on famine control, including two key earlier Ming texts on famine foods as well as close arguments by Xu himself on how to prevent and deal with famine. Xu's agronomic oeuvre coheres in a passionate project to restore the state, and in the first instance to protect the population from hunger and misery, not only because they trigger unrest and rebellions, but because such suffering is wrong.

This Confucian commitment manifests in Xu's attitude to *li*. Although we know from his letters that as a landlord Xu was fully capable of similar calculations of costs, losses and profits to Master Shen, and although the administrative sections of the *Complete treatise* are full of such quantitative details as costings of materials and wages for irrigation or land-reclamation schemes, when it comes to the type of financial details that would help a farmer decide which crop to choose, Xu is really not helpful. He must have had (or could have collected) detailed knowledge of prices and profit margins, not just in his native region of Shanghai but in the many parts of China where he had travelled.[54] Yet even though Xu includes a section of four chapters on commercial crops in the *Complete treatise*, his evaluations of their worth translate into such abstract phrases as:

> Tea is a marvellous plant: growing it brings abundant benefits, *libo*, and drinking it brings mental lucidity. Above princes and nobles trade it, and below the populace cannot do without it; truly it is both a commodity that is part of the everyday life of the people, and an aid to the revenues, *keli*, of the state.[55]

In a long section on another valuable cash-crop, the tallow tree, Xu tells us how much oil and how much tallow it is likely to yield in a good, medium or bad year, and how many scruples of beeswax should be added to ten pounds of the tallow to prevent the candles dripping – but when it comes to the *li*, the advantage or profit of planting this tree, he offers no hint of the price the oil might fetch. Its *li*, he says, is a hundred times that of other trees – by which Xu means that it provides lamp oil without encroaching on grain land.[56]

For Xu, as for most of his fellow officials, the concept of *li* could not and should not be reduced to financial profit. The social contract between state and *min* required officials to strive to bring prosperity and security to the people under their care, and this included instructing the *min* in beneficial, *li*, methods: disseminating useful crops, conserving the soil, improving water control and teaching better farming methods. In this conception of a healthy economy the key unit was the peasant

54 And see his arguments about comparative advantage in cotton production, discussed in Chapter 6.

55 *NZQS*: 1096–1097.

56 *NZQS*: 1065, 1068.

farm, producing enough for comfortable subsistence and the payment of taxes, with a small surplus that would provide enough cash for some creature comforts, and for decent weddings and funerals. At one level Xu and his contemporaries recognized that even small peasants' choices were inevitably affected by commercial concerns. Yet even here any mention of investment costs or profit margins remains at the abstract level. For Xu, the basic meaning of *li* as applied to a plant is that it is reliable and flexible in its uses, so that it can be effectively integrated into a farming system where famine is an ever-present threat.[57]

Xu Guangqi's remarkable empiricism, his experiments with plants and his study of history and geography for deducing patterns of locust attacks, were impelled by his perception of the pressing problems of the time – and of the levels at which to solve them. Contemporaries such as Master Shen tried to resolve these problems at the level of the landowner, Xu addressed them on the basis of a collaboration between state, officials and *min*. His experiments cannot be understood disembedded from his statecraft, even though what he wrote at that social level of action may seem less impressive and less innovative to us today than his field experiments and observations. Paradoxically, although Xu wrote on agriculture in the persona of a public servant, he was able to experiment successfully with plants and techniques precisely because he did not serve full-time as an administrator, but retreated from public life to his farm in Shanghai or his estate in Tianjin for long periods of leave.

What makes the *Complete treatise* unique, and very much a product of the late Ming, is not only Xu's remarkable experiments and deductions, but also his insistence that such knowledge only acquired meaning through the articulation of different levels of action. Xu's experiments with plants and techniques cannot be disembedded from a tiered framework of local administration and state-level policy, and an overarching philosophy of how to guarantee the welfare of the people. For Xu and his contemporaries, the *Complete treatise* represented the practical working-out of a political philosophy. It is because Xu wrote in this idiom that his incomplete manuscripts on agriculture were snatched up by a group of ardent young patriots searching for ways to serve the dynasty, and why in those tumultuous years of military defeats and civil collapse they found time not only to edit and publish the work but to present it to the emperor, who gave it his approval in the dying days of the dynasty, in 1639. The Ming fell in 1644. The new Qing rulers did not patronize works by Ming loyalists, among whom they counted Xu Guangqi: no further editions of the *Complete treatise* were published until 1837. Yet its value as a public *nongshu* could not be ignored. Shorn of the sections on restoring the polity, the technical sections of the *Complete treatise* were included

57 Bray and Métailié (2001).

in the imperial compendium of 1743, the *Shoushi tongkao* (Compendium of works and days), and thus were widely known throughout Qing China.[58]

Domestic virtues

As the fundamental occupation, *benye*, agriculture was an honourable and virtuous occupation – or preoccupation – for any man of any class. The authors or compilers of agricultural treatises, *nongshu*, included high functionaries on imperial commissions (as in the case of the Yuan *Nongsang jiyao* or the Qing *Shoushi tongkao*), individual officials (including Wang Zhen and Xu Guangqi) and educated landowners (such as Master Shen or Pu Songling). A great deal of the technical content of *nongshu* formed a pool of common agronomic knowledge and concern, some of which transferred readily between official and private *nongshu*. Yet there were many types or formulations of knowledge that did not cross the boundaries so easily. The context in which agronomic knowledge was embedded, and the values attached to it, differed depending on whether it was mobilized as a tool of statecraft or a technology of family success.

Official and private *nongshu* differed in their scope and their preferred layout, in their use of sources, in medium (printing versus manuscript, illustrations versus pure text), in the kinds of knowledge they sought to convey (general versus local), and in their intended readership (fellow officials versus fellow landlords). Private *nongshu* were by nature more local and ephemeral, and far fewer have survived as texts than public *nongshu*. Yet the titles of some hundreds have survived, recorded in bibliophiles' catalogues, dynastic bibliographies or references by contemporaries, confirming that the authorship of *nongshu* conferred social prestige. Indeed, the authors of both genres were equally convinced that they were contributing to the common good.

To officials, agronomy was a key technology of statecraft, one of a set of material techniques, together with building and public works, defence, hydraulics, transport, price regulation and famine-relief that officials routinely applied in the course of their duties. Beyond its immediate pragmatic importance, agronomic competence underpinned the moral order of society, state and cosmos. In the moral context of *quannong*, an individual who authored an official *nongshu* demonstrated both his agronomic virtuosity and his dedication to the welfare of the dynasty and its subjects. An emperor who commissioned or published such a work promoted the legitimacy of his rule through this benevolent project.

The authors of private *nongshu* typically portrayed their works as useful guides for kin and neighbours; prefaces or introductions generally present the work as

58 See Will (1994) on the role of the *Shoushi tongkao* in circulating knowledge, and Liang (1963: 105–109) on the editions of the *Nongzheng quanshu*.

beneficial beyond the family estate, contributing to the public wealth but on a local scale. The knowledge was offered as a guide to running a farm smoothly and productively, maintaining household income, managing tenants and workforce. Here farming and management skills fitted within a set of domestic technologies, also including ritual, geomancy and medicine, deemed essential knowledge for a family head in late imperial times. Together these skills formed an ethical-technical knowledge cluster focused on ritual and social propriety, family well-being and the perpetuation of the lineage and its patrimonial property.

8

AGRICULTURAL ILLUSTRATIONS

Blueprint or icon?

> The images, *tu*, are the warp threads and the written words, *shu*, are the weft. As warp and weft alternate to form the pattern of a fabric, *wen*, [so images and written words alternate to form the meaning of a text, *wen*] . . . To see the writing without the image is like hearing a voice without seeing the form; to see the image without the writing is like seeing a person but not hearing his words.[1]

The Song dynasty saw the rise of a new form for conveying technical knowledge, the illustrated register, *tupu*, in which graphic images were paired with explanatory texts. The category of *tu* encompassed a range of forms that today we might separate into pictures, maps or diagrams, a spectrum of visual forms collectively considered in the terms of the time as effective tools for communicating technical knowledge.[2] Intellectuals used *tu* to convey profound levels of understanding that went beyond words: an example would be the diagrams devised by philosophers to develop cosmological arguments or textual exegesis.[3] Other forms of *tu*, often more straightforwardly illustrative, served to convey knowledge and understanding to the uneducated and illiterate. Here the rationale was that *tu* conveyed the essentials, *yao*, simply and accessibly.

Technical *tu* had a long and rich history in China even before the rise of print culture in the Song. As the printing industry developed and diversified, facilitating

1 *Tupu lüe*: 837a.

2 For an introduction to specific categories of *tu*, their varieties and uses, see the papers in Bray *et al.* (2007); Bray (2007a) offers a general introduction to *tu* and their uses over time.

3 Lackner (2000).

the accurate reproduction and mass circulation of images, *tu* became increasingly common, eventually permeating the visual culture.[4] The educated elite began to make systematic use of *tu* to record technical, often technological, information both for each other, and to transmit information and ideas to ordinary people. *Tu* were enthusiastically adopted by late imperial statesmen and rulers as a tool of governance and pedagogy, including for the promotion of agriculture, *quannong*. This chapter traces the history of agricultural illustration in late imperial China, to ask what forms it took, what purposes it served, what it accomplished, and how or whether it developed over time.

Historians of Western science and technology have argued that the development of technical drawing played a crucial role in the rise of modern Western science and society: the act of technical drawing stimulated practitioners to think creatively, while the development of drawing conventions encouraged precision of observation and accuracy of represention.[5] Unlike the post-Renaissance West, China did not develop a self-conscious, structured discipline of technical drawing that progressed hand in hand with steadily evolving disciplines of science.[6] Yet asking why agricultural *tu* took the forms they did, in which contexts they were used, and why their use continued even when from a Western perspective they appear obsolete or even misleading, suggests some interesting insights into the goals that the Chinese state and producers of technical knowledge set for themselves.

The earliest Chinese technical writing on farming dates back at least as far as the late Warring States, and visual depictions of farming tasks and tools proliferated from the Han period.[7] Yet agricultural illustration – the deliberate pairing of agricultural text with images or graphics – seems to have begun no earlier than the Song dynasty. One work produced at the beginning of the Southern Song, Lou Shou's *Gengzhi tu* (Pictures of tilling and weaving),[8] and another produced

4 Historians of printing and of art usually see this as an effect of a second 'print revolution' in the mid-Ming, when cheap printed books and images began to circulate widely, with considerable effect on popular culture; see, for example, Clunas (1997); Hegel (1998).

5 See, for instance, Lefèvre (2004).

6 Golas (1999, 2001, 2007).

7 See the illustrations in Bray (1984) and Wang (1995).

8 Lou's given name can be pronounced Shou, Shu or Chou. Chou seems now to be standard among historians in China (e.g. Wang, 1995). Shu is preferred among some Western art historians, including Roslyn Hammers, whose magnificent study unfortunately appeared too late for me to engage with her arguments in my revision of this chapter, originally published in 2007. Hammers' understanding of the purpose of the *Gengzhi tu* broadly concurs with my own. She sees it as a proposal for reform structured at two levels, the first agronomic-material (illustrating sophisticated techniques), and the second metaphoric, in that each of the scenes and poems of Lou Shou's work evoked well-known precedents of wise rule or of reform that benefited the common people. The series as a whole thus sent a strong political message, stressing the dignity of the workers' labour, and their active role in contributing to the social order (Hammers, 2011).

in the mid-Yuan, Wang Zhen's *Nongshu* (Agricultural treatise), launched the two predominant Chinese traditions of agricultural illustration.

The *Gengzhi tu*, presented by Lou Shou to the emperor circa 1145, was originally designed as an album, a common artistic medium of the period consisting of a sequence of paintings on silk, each paired with a poem. It was subsequently reproduced in a range of media, including stone carvings, woodblock prints and painted porcelain. Wang Zhen's *Nongshu* of 1313 was in many respects a conventional, text-based agronomic treatise. However, it contains a long section entitled *Nongqi tupu* (Illustrated register of agricultural implements), which was a complete innovation: for each item of equipment a drawing, deliberately technical in style, completes the prose explanation. The work was clearly designed specifically for the medium of woodblock printing, and subsequent reproductions or adaptations were all in that same medium.

Wang Zhen intended his visual-verbal depictions of technical devices to serve as the equivalent of blueprints, allowing officials to introduce more advanced technology to backward regions. Lou Shou's work was intended to convey messages that were social and moral as much as technical: a primary concern was to foster a proper respect for the role that farming families played in sustaining the social order. Although in several cases Lou Shou's very detailed paintings show more technical detail than the more schematic renderings of Wang Zhen's *Illustrated register*, his more artistic choice of medium and style produced pictures that were closer to genre scenes than to blueprints.

From the perspective of how knowledge and governance are organized in the modern world, it might well appear that these two experiments with the visual rendering of technical knowledge should be treated as distinct endeavours. Yet despite their differences in conception and execution, the *Gengzhi tu* and the *Nongshu* shared a common goal. Both were both intended to serve the overarching pedagogical purpose of "promoting farming", *quannong*, a primary responsibility for anyone involved in imperial government. Exactly what place, then, did technical illustration play in this project?

The *Gengzhi tu* and its tradition: icons of the imperial order

The Southern Song scholar Zheng Qiao (1104–1162) completed his *Tongzhi* (Historical collections) in about 1150. The work contains a famous essay, the *Tupu lüe* (A brief account of the illustrated register), in which Zheng analyses the relations between text and image and the role of each in conveying knowledge.[9]

9 See the quotation that opens this chapter.

At the time that Zheng was writing, the popularization of woodblock printing had opened up new opportunities for writers on cosmology, mathematics, materia medica, building and other technical subjects to develop richer forms of communication through the use of diagrams, charts and illustrations. Zheng's essay refers to several of these fields, but not to agriculture. In his recent study of the history of agricultural illustration in China, Zhang Zhongge takes this omission to mean that no agricultural illustrations had yet been produced,[10] and certainly we have no evidence to suggest that the agricultural writers of Zheng Qiao's time, Chen Fu for example, had begun to use illustrations to convey their ideas.[11] But officials and emperors, it seems, were more adventurous. The notion of using images of farming for pedagogical purposes, whether or not paired with text, seems to have emerged during the later Song, and to have done so in the context of government.

Promoting agriculture, *quannong*, was one of the principal responsibilities of government in imperial China. All Chinese emperors and administrators spoke freely of their devotion to this cause, but the Northern Song emperors and their civil servants were unequalled in the scope and coordination of their practical efforts to help farmers and increase agricultural production. Dykes and irrigation works were built, marshes converted to farmland, seeds of improved varieties imported and distributed, and instruction was provided through local magistrates' offices in improved methods of crop rotation, sowing and fertilization.[12] In 1012, under the emperor Zhenzong, quick-ripening rices from Champa were introduced to the Yangzi Delta along with detailed instructions for their cultivation, such as those reproduced in the *Song huiyao gao* (Drafts for the administrative statutes of the Song dynasty).[13] Printing played an important role in the official circulation of technical information: both the sixth-century agricultural treatise *Qimin yaoshu* (Essential techniques for the common people) and the eighth-century farming calendar *Sishi zuanyao* (Compendium of essentials for the four seasons) were printed and distributed to local magistrates by the imperial printing office in 1020,[14] and descriptions of new crops and improved techniques were circulated to and through magistrates' offices on printed handouts. However, these printed texts do not seem to have been illustrated in any way.

10 Zhang (1995: 1).

11 Chen Fu was the author of a short work on Southern rice farming, entitled *Nongshu* (Agricultural treatise), prefaced 1149; see Chapter 2 and below.

12 See Chapters 2 and 6.

13 See Bray (1984: 492). Within a century Champa rices were said largely to have replaced traditional round-grain, long-season rices in much of the region, in part because they could be used in a double-cropping alternation with winter wheat (Amano, 1979: 106; Ho, 1956).

14 Commercial printed editions of both works had been available at least since 996 (Hu, 1985: 46).

Another strategy for promoting agriculture and sericulture, also apparently initiated in the Northern Song, was to adorn the walls of palaces, prefectural offices or other public buildings with images of farming activities. Writing about a century later, the scholar Wang Yinglin noted that 'the first pictures of peasant households engaged in farming and weaving appeared on the walls of the Yanchun Pavilion [part of the Imperial compound] during the Baoyuan reign-period (1038–40) of Emperor Renzong'.[15] The artists who produced these murals must have been humble artisan painters, for if the images had been designed by a scholar-official his name would certainly have been recorded.

Since the Yanchun Pavilion murals were located in the imperial palace they were obviously not intended to instruct farmers directly in improved techniques. At one level their role was certainly iconic: commissioning paintings of the production of grain and silk for the palace represented imperial recognition of the importance of these activities.[16] But they may also have been intended to introduce the women of the court to sericultural skills. Gaozong (r. 1127–1163), the first emperor of the Southern Song, referred to the Yanchun Pavilion murals as a precedent when he declared that 'rearing silkworms should be done in the imperial palace, so that all will know the hardships of farm work'.[17]

The Yanchun Pavilion paintings must have depicted practices typical of North China, since at that time the northern plains were still both the political and the economic heartlands of the empire. When the northern provinces were conquered by the Jurchen Jin dynasty in 1126, the grainlands of the Yellow River plains and the main silk-producing region of Shandong were both lost to the Song state. Sericulture had been practised in the Lower Yangzi for some centuries, but techniques and equipment only started to catch up with the advanced methods of Shandong and Sichuan in the eleventh and twelfth centuries.[18] The rice fields of Jiangnan were fertile, but with the loss of the north they were required to feed a population swollen by refugees. Immediately after his accession to the throne, the first Southern Song emperor Gaozong issued an edict calling on the people to return to the land; as just mentioned, he also declared that sericulture should be practised at court.

15 *Kunxue jiwen* (Study notes), quoted in Wang (1995: Preface, 1).

16 As did the annual spring ceremony where the emperor himself ploughed a ritual furrow in the royal fields near the capital (Figure 8.4), while all across the realm local magistrates were supposed to do the same. The European equivalent was the traditional blessing of the plough by Christian priests, a ceremony whose pagan origins survived until recently around England in the festivities of Plough Monday.

17 Quoted in Wang (1995: 33, 35). Three attempts were also made to revive the legendary court sacrifices to the silkworm deity during the Song dynasty, in 1111 before the fall of the Northern Song, and in 1132 and 1145 just after the court had been established in Hangzhou, but all failed (Kuhn, 1988: 255–256).

18 See Kuhn (1987: 163–173).

It is in this political context that Lou Shou (1090–1162) produced his paintings and poems, completed six or seven years after the fall of the North.[19] Lou Shou was a native of what is now the Zhejiang province. He came from a distinguished family in the Ningbo region, and followed his father into official service. In the course of his life he had appointments all over China, with postings at the capital and as far south as Guangzhou, but in the early 1130s he was magistrate of Yuqian County, in the heart of the Jiangnan region where rice farming techniques were most advanced. It was in Yuqian, in 1132–1134, that Lou Shou completed the *Gengzhi tu*, a set of 45 paintings showing the main tasks of rice farming and sericulture, each inscribed with a poem of the author's own composition.[20] The *Gengzhi tu* was presented to the emperor in 1153 or 1154.[21]

In his introduction to the first printed edition, Lou Shou's nephew Lou Yue writes:

> Bofu (Lou Shou) was genuinely concerned with the common people's lives. He felt deep sympathy for the hard toil of farmers and women weavers, and investigated every aspect of their tasks for his two series of pictures on ploughing [rice farming] and weaving [sericulture]. For farming he depicts every step from soaking the seeds to storing the grain in the granary, twenty-one steps in all; for sericulture there are twenty-four steps, from washing the eggs to tailoring the cloth. Each step is illustrated with a picture accompanied by an eight-line, five-character poem. All the tasks of farming and sericulture are minutely and clearly portrayed. Although local practices vary in their details, all the main principles are correctly set out here, as everyone who sees it will recognize. Soon after the work was completed, an imperial envoy on his tour of inspection gave a report of it to the emperor who, on the advice of his trusted ministers, summoned Lou Shou for an audience. The paintings thus came to court. The emperor praised and rewarded Lou, and gave orders for the scenes to be painted on the screen of the Inner Court, with Lou Shou's name on them.[22]

Lou Shou's innovation was to bring together technical pictures and poems on agriculture and sericulture, apparently for the first time. It was a long-standing custom for scholar artists to inscribe their paintings with poems – calligraphy, poetry and painting were considered part of the same creative process. One common form

19 Key works on Lou Shou and on the *Gengzhi tu* include Franke (1913), Pelliot (1913), Kuhn (1976), Watabe (1986), Wang (1995) and Hammers (2011).

20 Yuqian is the modern Lin'an county in Zhejiang province; it lies about 40 km west of Hangzhou, up the Qiantang River.

21 Watabe (1986: 4–5), Wang (1995: 33 ff).

22 *Gongkui ji* (Collected works of Gongkui), quoted in Wang (1995: 33, 36).

of composition was the album, containing a sequence of poems and paintings on a single theme. Most albums were intended for display to other members of the elite: the poems were sophisticated and the calligraphy cursive. Yet in their simpler forms, images and poems were both considered effective media for conveying knowledge to the uneducated (peasants, women or children). As Zheng Qiao put it in his essay, unlike text an image was easily apprehended and conveyed the essentials, *yao*, at a glance.[23] Likewise, rhyme and rhythm rendered the simpler ballad forms of poetry more memorable than prose. Several distinguished Song statesmen-scholars had written poems in ballad style to disseminate such useful information to the uneducated as moral precepts or basic medical prescriptions for common ailments. The statesman Wang Anshi, notable for his farm policies, had written a set of fifteen short ballads on farm tools.[24] Meanwhile, various elite artists during the Song enjoyed painting scenes from popular life in the city or the country, and their works often included detailed depictions of technical artefacts such as water mills or boats, or of technical processes such as spinning hemp or silk.[25] But Lou Shou seems to have been the first person to put technical pictures and poems together into what was essentially an album. Should we conclude that what Lou Shou had in mind in designing the *Gengzhi tu* was a work that could be used to communicate at both elite and public level?

Lou Yue claims that 'the pictures were so vivid, and the poems so expressive, that at court and through the country they were passed around and learned by heart'.[26] The Yuan writer Yu Ji (1272–1348) says: 'In the previous dynasty [i.e. the Song] pictures of farming and sericulture, *gengzhi tu*, were painted on both the east and west walls of the entrances to district and prefectural offices for the common people to gaze upon, though now they are very rare'.[27] There is some ambiguity in this last statement, where the term *gengzhi tu* could refer either to copies of, or to works directly inspired by Lou Shou's series, or to other depictions of farming and sericulture. Should we presume that the murals of "sericulture and wheat", *canmai tu*, that the Yuan emperor Yingzong ordered for his palace in 1322 were inspired by Lou Shou's work but adapted to northern conditions, or should we think of Lou Shou's *Gengzhi tu* as one particularly sophisticated contribution to a broader tradition of civic representations of farming?

There is doubtless some exaggeration to Lou Yue's statement, which suggests that Lou Shou's work soon became known in every corner of the realm. But even

23 *Tupu lüe*: 837a.

24 Translated in Hammers (2011: 227–230).

25 For examples of paintings of mills and textile processes, see Kuhn (1987: Chapter 3).

26 *Gongkui ji*, quoted in Wang (1995: Preface, 1).

27 *Daoyuan xuegulu*, (Daoyuan's notes on antiquity), quoted in Watabe (1986: 3) and Zhang (1995: Preface, 1).

though none of the earliest versions has survived, the work did quickly achieve wide diffusion, and it does seem very soon to have entered the cultural repertory as *the* way to depict rural livelihoods. Lou Shou made a copy of the *Gengzhi tu* for himself and presented the original to the emperor in 1153 or 1154, by which time he had been transferred to Yangzhou.[28] The palace fresco was copied from Lou's original, and it seems that two artists, Liang Kai and Liu Songnian, both painted copies in the 1190s. Lou Shou kept the copy he himself had made, and it was used by his grandsons Lou Hong and Lou Shen for an engraving on stone in 1210. Sometime in the 1220s, according to Lou Shou's great grandson Lou Shao, Wang Gang, the prefect of Shaoxing, had the first woodblock engravings made of Lou Shou's own copy, in order, he said, to show people who were well-fed and warmly clothed where these comforts came from. This printed edition seems to have circulated quite widely in the late Song; in 1237 Lou Shao wrote the preface to a new edition.[29]

Neither the original scrolls nor the direct copies in the palace or engraved on stone by Lou Shou's grandsons have survived. It is believed that a scroll of sericultural scenes, now in the Heilongjiang Provincial Museum, is a Song copy made directly from the original scroll that Lou Shou presented to the emperor by the Imperial Art Academy. Fragments of Liang Kai's copy (also depicting sericultural scenes) are now in the Cleveland Museum. In the early years of the Yuan dynasty, the scholar-painter Cheng Qi also painted a version, based apparently on Wang Gang's woodblock version (which in turn was based on the copy Lou Shou himself had made and kept). This version is now in the Freer Gallery in Washington.[30]

During the early Ming dynasty, Wang Gang's woodblock version was reproduced in the *Yongle dadian* (Yongle encyclopedia) of 1407, and reissued by an official named Song Zonglu in 1462.[31] The latter version was extremely influential because copies found their way to Japan and Korea, where they inspired many local variants, including a version by the famous genre painter Kim Hongdo (1745–?), preserved in the National Museum in Seoul.[32]

Song Zonglu's woodblock *Gengzhi tu* also provided the model for a considerably reworked version that figures at the beginning of the *Bianmin tuzuan* (Illustrated collection for the convenience of ordinary people), a household handbook whose author is said to have been called Guang Fan. Like many other popular household manuals, the *Bianmin tuzuan* contains basic practical instructions on farming, textile

28 Watabe (1986: 5).

29 Watabe (1986: 6–7).

30 Watabe (1986: 16, Table 3); Wang (1995: 37). The Freer Gallery pictures are available online, together with detailed documentation (Freer Gallery of Art, 2010a, 2010b).

31 This edition of the *Gengzhi tu* is the one reproduced in Franke (1913), along with the version published in the *Shoushi tongkao* of 1742 (see below).

32 Watabe (1986: 18–20) on Korea; Watabe (1986: 20–28) on Japan.

production and building procedures; advice on divination, ritual and etiquette; tips for household maintenance; recipes for dishes and medical prescriptions. The earliest edition of the *Bianmin tuzuan*, dated 1502, has been lost; the illustrations to the surviving Jiajing edition of 1544 are quite rough, but those to the Wanli edition of 1593 are considered 'exquisite'.[33] Under the title *Nongwu zhi tu* (Pictures of farming tasks) and *Nühong zhi tu* (Pictures of women's work), the author offers us 15 scenes of rice farming and 16 of sericulture, accompanied by a set of four-line seven-character poems, which he tells us he composed in the Wu dialect used by local farmers (Figure 8.1).[34]

The printing industry expanded rapidly in the mid-Ming, bringing book prices down and supporting an explosion of popular publications, among which

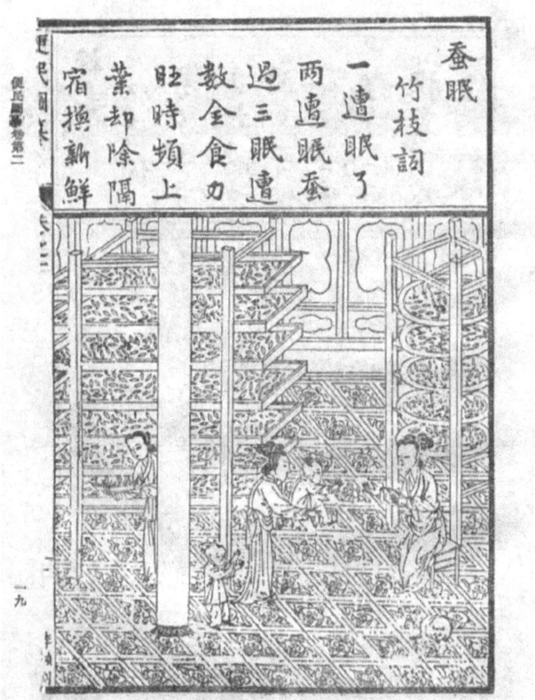

FIGURE 8.1 'Silkworms' sleep', *Bianmin tuzuan*, 1593 edition. The poem notes that the silkworms sleep three times, offering welcome breaks from the continuous toil of feeding and cleaning.

33 Wang 1995: 66; these are the pictures reproduced in the modern edition of 1959 by Shi Shenghan and Kang Chengyi, though the text is the Jiajing edition.

34 *BMTZ*: 1.

such household handbooks figured prominently. Publishers of the period added illustrations to all kinds of books to help sales, and the *Gengzhi tu* illustrations appear to have been considered apt for works aimed at landowning households. Another popular household manual of 1607, the *Bianyong xuehai qunyu* (Seas of knowledge and mines of jade: encyclopedia for convenient use) contains similar woodblock pictures and identical poems to the *Bianmin tuzuan*, though in several cases poem and picture are combined differently.[35] A new version of the *Bianmin tuzuan*, published by Kanō Einō in Japan in 1676, served as the inspiration for various distinctively Japanese illustrations of farming and textile work, included as separate sections in the farming or household manuals that proliferated during the Edo period.[36]

The *Gengzhi tu* series gained its greatest official prestige during the Qing dynasty, when no fewer than three emperors, Kangxi (r. 1662–1722), Yongzheng (r. 1723–1735) and Qianlong (r. 1736–1796), commissioned new painted versions or stone carvings, composed additional sets of poems to accompany the pictures and had the new versions printed in woodblock format by the imperial printing house. As foreign rulers, the Qing emperors were especially anxious to show themselves to be good Confucians and ardent supporters of the Chinese moral order, committed to promoting agriculture, *quannong*. Patronage of the *Gengzhi tu* was an ideal instrument for this ideological task.

The Kangxi emperor commissioned Jiao Bingzhen, a court artist, to paint a version of the *Gengzhi tu* using Western perspective. Jiao presented his completed work, consisting of two sets of 23 scenes, to the emperor in 1696. Kangxi was delighted with it, and wrote a preface in which he reminded his readers of the hardships suffered by the common people in order to produce the grain and cloth they took for granted. He inscribed each picture with his own seven-character four-line poems, then had the work 'engraved, printed and distributed among his subjects'. The engraver who produced the woodblocks was Zhu Gui.[37] Jiao's pictures were copied again, with slight variants, on the order of the Yongzheng emperor, who wrote an eight-line, five-character poem to go with each of 46 scenes, substituted for Lou Shou's original poems. The Qianlong emperor later wrote his own set of poems for these scenes.[38] The images were also included in the compilation of agricultural knowledge entitled *Shoushi tongkao* (Compendium of works and days), commissioned by the Qianlong emperor and completed in 1742. Finally, Jiao Bingzhen's images were introduced to Japan in the early nineteenth century, and served as a stimulus for many local versions.[39]

35 Kuhn (1976) reproduces this series in its entirety. Kuhn (1976: 344–349) and Watabe (1986: 13) both provide tables comparing the contents of various *Gengzhi tu* series.

36 Two fine examples are illustrated by Watabe (1986: Figures 33a–o, 34a–k).

37 Wang (1995: 79–80); Kuhn (1976: 339). Kangxi's poems are given in Wang (1995: 200–201).

38 Wang (1995: 205–209, 213–215).

39 Watabe (1986: 20–28).

The Qianlong emperor also acquired Cheng Qi's paintings, dating from the Yuan dynasty and based directly on Lou Shou's original. He was so fond of them that he wrote his own poems in running script next to Lou Shou's, using the same rhymes. In 1769 he had the Imperial Art Academy make line copies of Cheng Qi's paintings. The drawings, together with the emperor's poems, were carved on stone slabs and kept in the Yuanmingyuan; half of them were destroyed in 1860 when Anglo-French forces burned the Summer Palace, but 23 survived, 12 of farming, 8 of sericulture and 3 'eroded beyond recognition'. They are now in the China Historical Museum.[40] Another version on a single giant stela of Kunlun stone, 2 m high, 1 m wide and 0.7 m thick, was set up in the quarter of Beijing where the imperial silk factories were located; it has survived to the present day. According to Wang Chaosheng and his colleagues, this stela was erected no later than 1753.[41]

Lou Shou's *Gengzhi tu* was not only prized by officials and incorporated into accessible handbooks. Its imagery quickly entered the craftsman's repertory of popular scenes. The earliest surviving example of the use of the *Gengzhi tu* as a decorative motif is a round fan, dating from sometime in the late Song, which shows all the tasks of rice farming set out in a single landscape.[42] Later, scenes from the *Gengzhi tu* were used to decorate ceramics and lacquer screens, including works intended for the European export market, and they also figured in woodblock New Year prints.[43] Often several tasks shown individually in Lou Shou's paintings are grouped together in a single scene or landscape in these other media, as they are also in various Korean and Japanese paintings on the theme, and many depict local tools, costumes, crops and divisions of labour.[44] An eighteenth-century Chinese painted wallpaper in an Austrian castle shows scenes of rice farming and sericulture closely resembling the 1696 version of the *Gengzhi tu*, intermingled with scenes of tea cultivation and pottery production. The general effect is very picturesque, but only a few of the

40 Wang (1995: 128). Rubbings of the complete series were published by Pelliot in 1913; see also Wang (1995: 133–148).

41 Watabe (1986: 10–11); Wang (1995: 126–129, 209–212). On the stela, see Wang (1995: 129ff).

42 Watabe (1986: Figure 4); the fan is in the Beijing Palace collection.

43 Franke illustrates several pieces from German collections of the period. They include three fine porcelain plates from the Kangxi period, each showing a single scene from the rice farming series, and two vases, also Kangxi, one showing various rice farming scenes scattered across a single landscape, the other showing a sericultural scene (Franke, 1913: Plates I, II and III, facing pp. 80, 88 and 96). Watabe (1986: 17–18, Figures 19, 20) discusses three of the same pieces, and also shows a late Qing New Year print from Suzhou modelled on Jiao Bingzhen's painting (Figure 21). Watabe also includes several examples of Japanese screens based on versions of the *Gengzhi tu*. A Chinese twelve-panel "Coromandel" lacquer screen from about 1700, showing a rural landscape containing scenes from the *Gengzhi tu* on one side, and a busy prefect's court on the other, was offered for sale in 2003 by a London antique dealer (Marchant, 2003: 64–75).

44 For example, the Korean two-panel painting of men and women in Korean dress going about "the happy tasks of farming families" shown by Watabe (1986: Figure 24), or the 1882 carved panels from a Henan farmhouse, discovered in 1978 (Wang, 1995: 159–160). A Japanese woodblock version from

stages depicted in the *Gengzhi tu* are shown, and the painters have omitted such key technical details as the cog-wheel of the chain-pump.[45]

Today there are still some 60 editions and variants of the *Gengzhi tu* in existence throughout East Asia. What accounts for its long-lasting popularity in late imperial China, Korea and Japan? It is often presumed by historians of technology that the work was intended to provide advanced technical knowledge that could be used to improve agricultural and textile practices. Certainly it was produced at a time when state officials were putting great energy into such development efforts. And certainly Lou Shou's elegant pictures depict in vivid detail the practices of growing rice and of making silk that were typical of one of China's technically most advanced regions. Yet the images of equipment are often incomplete from the point of view of anyone trying to reproduce their structure: instead of focusing on the tools and machines, or on the hands of the workers, the artist often chooses to foreground rustic landscapes of willows and waterways. He dwells affectionately on thatched cottages and distant mountains, an old man looking on from the bank of the rice-field, a toddler playing with a dog in front of the silk-reeling machine, one woman suckling her infant and another sewing as they sit in front of the stands holding the silkworm trays. In Lou Shou's pictures the loom is half-hidden behind a wall, the plough is buried in the mud. His ballad-like poems emphasize mood and morale more than technical precision:

> A clear frosty day
> When a strong wind tears the leaves from the trees
> Is best for threshing.
> The irregular beat of the flails rings out,
> The brown hens peck at the fallen grain
> While the crows chatter gleefully.
> Back home the farmers shake off the dirt and dust,
> And in their cottages these nights they burn the stalks of the grain.[46]

Clearly Lou Shou intended his pairing of poems and images as a powerful medium for conveying an important message. Perhaps he chose the form so it should be accessible to humble and uneducated people as well as officials and scholars. But what exactly was the message?

the late seventeenth-century shows women transplanting and weeding the rice (*Nōgyō zensho*: 1/2b). Although women did participate in such tasks in some parts of southern China, the conventional Chinese view was that all field tasks were men's work, and that women's work was confined to the home. When women are shown in the farming section of Chinese versions of the *Gengzhi tu*, they are usually bringing baskets of food and drink to the men working in the fields. One exception was cotton-picking, represented in the *Mianhua tu* of 1765 as a task for women and children (Figure 2.4).

45 Berger *et al.* (1996: 102).

46 Tr. Bray (1984: 357).

The *Gengzhi tu* depicts scenes of production, but although it does contain many telling technical details, taken in conjunction with the poems I would argue that it is as much the production of social harmony and political order that is at stake as material production. When Lou Shou painted the *Gengzhi tu* the Chinese empire had just suffered a cataclysmic defeat. The imperial state had lost its traditional economic base in the northern plains, and now had to recognize its dependence on the rice regions of the south. In fact the contribution of the south to the national exchequer had been increasing steadily for three hundred years, but this growing importance had gone unmarked in the technical literature: all the classic farming treatises concentrated on the dry-land farming systems of the north. Furthermore, it had been common to underrate the skills and hard work of southern farmers: the south was commonly referred to as "the land of fish and rice", implying that it was such a naturally productive region that bounteous harvests sprang up unaided. The *Gengzhi tu* was the first work to celebrate the indispensable role of the irrigated Jiangnan landscape in supporting the imperial order, and to show the skills, hard work and technical sophistication that made this contribution possible. Together Lou Shou's pictures and poems insist upon the daily toil that went into making the land yield such riches; they movingly portray the sweat, dedication and sacrifice that went into producing the rice and silk with which the peasants paid their taxes. They also emphasize the complexity and sophistication of the processes involved, but frequently as an object of marvel and admiration rather than a process to be explained. In the poem on the drawloom, for example, Lou Shou writes: 'hand and mind work in obscure resonance, before the eyes a colourful pattern grows'.[47]

Lou Shou's images emphasize working skills and bodies in motion, sometimes showing several stages of one basic process (winnowing and sifting the grain, for example).[48] One Korean version of the reaping scene actually shows the details of a tool in action. The sickle has a long wooden hook at the back of the haft, slightly below the blade, and the reaping takes place in three steps: first the reaper uses the hook to gather a bundle of rice stems into his left hand; then he reverses the sickle to cut the bundle with the metal blade; finally he lays the bundle on the ground, to be collected into a sheaf (Figure 8.2).[49]

47 See Wang (1995: 191); *Mohua*.

48 As do the illustrations to the late Ming work on crafts and production, Song Yingxing's *Tiangong kaiwu* of 1637; see Golas (2007).

49 My thanks to François Sigaut for providing me with a photocopy of this version of the *Gengzhi tu*, which he found in Japan, and for pointing out that the reaping picture shows the various steps of cutting the rice in sequence. The version in question, entitled *Nongjia fengsu tu*, is Korean (Watabe Takeshi, pers. comm. via Charlotte von Verschuer, 2003). Although hooked sickles are quite common around East and Southeast Asia, I have never seen them depicted in any other pre-modern East Asian agricultural works or pictures. All the other versions of the *Gengzhi tu*, whether Chinese, Japanese or Korean, show simple sickles with small metal blades set at the top of a straight wooden shaft. Wang Zhen's *Nongqi tupu* shows a variety of sickles, billhooks and a scythe, but no sickles of this type.

FIGURE 8.2 Reaping scene, *Gengzhi tu*, eighteenth-century Korean edition.

More than the precise details of technical action, however, I would say that the *Gengzhi tu* pictures and poems were intended to emphasize toil and effort, and that later versions share this concern. In the scene of irrigating the growing crops, Cheng Qi's copy of Lou Shou's painting shows a team of four people working a chain-pump in the foreground, and a man using the slow and old-fashioned bucket on a sweep in the background (Figure 0.1). Lou's poem praises the advantages of the chain-pump, which he describes as so efficient that it allows the irrigation team to return home before sunset. By the mid-Ming, however, Guang Fan's poem on the same chain-pump laments 'Sore legs, aching back, busy day and night'.[50] Two centuries later, Jiao Bingzhen does not foreground the chain-pump, although – or more likely because – it was the most common irrigation device in South China. Instead Jiao shows us a scene of fields around a lake, with a chain-pump at work

50 *BMTZ*: 10. The illustration is closely modelled on the Cheng Qi version except that it omits the swape and bucket and shows only the chain-pump. Hammers (2005: 135) argues that the earlier version shows a woman pedalling the chain-pump along with three men, underscoring the message in the preceding poem, which emphasizes how easy the pump is to use. In the *Bianmin tuzuan* illustration the woman has vanished.

half visible behind a tree; in the foreground a man is lifting water into a field with an old-fashioned bucket on a sweep (Figure 8.3). Wang Chaosheng believes Jiao's picture corresponds to irrigation practices in northern China (he apparently has not noticed the pump in the background).[51] I see the prominence Jiao gives to the slow and clumsy sweep as a metaphor for the many tedious hours of hard work needed to water the crops whatever instrument is used.

Lou Shou's paintings evoked the village roots of imperial wealth and power; his poems, echoed in the later prefaces by Wang Gang and Kangxi, reminded the ruling class that these goods were produced at a cost and should not be abused. For officials, the *Gengzhi tu* depicted an ideal social contract in which the peasantry toiled to produce the rice and silk upon which the imperial government depended, and the emperor and his officials beholding the scrolls, or the landlords reading the books portraying these scenes, were reminded that they should be grateful for the sacrifices entailed and devote themselves to promoting the welfare of the people, or of their tenants.

FIGURE 8.3 Irrigating the fields, *Gengzhi tu*, 1742 woodblock version based on Jiao Bingzhen's painting of 1696.

51 Wang (1995: 77).

From the perspective of imperial rule, promoting agriculture, *quannong*, was both a material and an ideological or cosmic endeavour. The original *Gengzhi tu* certainly combined both dimensions, offering technical observation as well as charming scenes of rural life. But its greatest success was as an icon celebrating a fruitful social harmony. Of paramount importance was the interdependence of the state and the peasantry. In pairing rice farming and sericulture, the production of food and of clothes, the work also underlined the essential complementarity between male and female work. The interdependence of generations is another key theme: mature adults are shown at work, while old people look on or help out by minding the children (see Figures 2.1 and 8.1). The central theme of families working together to achieve abundance is certainly one reason why the scenes of the *Gengzhi tu* were so popular as decorative motifs in the late Ming and Qing: they ranked with vignettes of a son's success in the imperial examinations, or depictions of rosy-cheeked children at play, as auspicious images of happiness, virtue and harmony.

When the Qing emperors took the *Gengzhi tu* up with such enthusiasm, it must surely have been for its iconic rather than its technical value. Certainly there were few dramatic transformations in rice farming equipment between the Song and the Qing, but there had been several significant technical improvements in farming practice, notably in irrigation, and in fertilizing materials and methods.[52] Yet the new versions of the *Gengzhi tu* commissioned by the Qing emperors reflect nothing of this, either in the scenes shown or in the poems. The late Ming *Bianmin tuzuan* version, which was intended for landowners rather than courtiers, mentions in one poem the use of bean-cake for fertilizer, which by that time was a routine practice. But no mention of the new commercial fertilizers is made in any of the new poems written by the Qing emperors.[53] However, while Song and Yuan versions ended the rice farming sequence with the filling of the granary, the version that Jiao Bingzhen produced for the Kangxi emperor adds a new scene: a solemn thanksgiving ceremony at the family altar (see Figure 0.2).[54] This is by no means the only example of Qing rulers favouring symbolic over material instrumentality. Some 50 years later, the huge compendium *Shoushi tongkao* commissioned by the Qianlong emperor would devote more space to rituals and to "auspicious grains", *jiahe*, than to commercial crops or flood control.[55]

52 Li Bozhong (1998a, 1998b).

53 *BMTZ:* 6; Wang (1995: 200–201, 205, 209).

54 The *Bianmin tuzuan* had added a picture of a riotous harvest feast with drunken tricks, dancing and much vomiting; though the festive table is set before the landlord's family altar, only the very edge of the altar is visible; *BMTZ*: 16 (Figure 0.3).

55 Bray (1984: 72–74). Ears of grain, or indeed any products of the soil, that were especially large or formed in curious and auspicious shapes were believed to be a sign of Heaven's favour, and they were supposed to be taken to the local magistrate who would send them to the Emperor. For a fascinating account of how such lucky omens (in this case, a specially fine melon) were interpreted by the imperial ruling class, see Schneewind (2006).

Wang Zhen's *Illustrated register*: woodblock printing and the pairing of prose and image

By the time that Wang Zhen composed his *Nongshu* (Agricultural treatise) at the beginning of the fourteenth century, woodblock printing was highly developed, and Wang took full advantage of the opportunities this medium offered for expressing knowledge not only through written text, but also through diagrams and illustrations. Wang Zhen's linking of text and graphic follows a completely different logic from that of Lou Shou, offering an early example of what we understand today by technical illustration.

Wang Zhen's *Nongshu* consists of three sections, one consisting of general instructions on farming and textile methods (*Nongsang tongjue*), one on crops (*Baigu pu*, The hundred grains), and one (*Nongqi tupu*, Illustrated register of agricultural implements) on tools, equipment and machinery. Wang Zhen dated his preface to the *Nongshu* in 1313, and it seems that two woodblock editions may have been printed during the Yuan dynasty, one at that date (missing the section on crops) and another (including all three sections) a few years later. Unfortunately, neither has survived. The numerous later editions fall into two main lineages. One derives from the version published in the *Yongle dadian* (Yongle encyclopedia) of 1407, possibly based on a manuscript copy; although no copy of the Yongle edition has survived to the present, it was reproduced in the imperial collection *Siku quanshu* of 1782. Another derives from a woodblock edition of 1530 (during the Jiajing reign period), published by the provincial governor of Shandong from an engraving of a manuscript version.[56]

In the bibliographical introduction to his critical edition of the *Nongshu*, Wang Yuhu explains the principal textual differences between the two lineages, which are significant in terms of the ordering of sections and the numbering of chapters, but otherwise rather minor. Wang notes that the illustrations to the Jiajing edition of 1530 give more prominence to the implements or equipment, which are often simply presented without any background, while the *Siku quanshu* edition tends to present them in a more elaborate setting, with more emphasis on human figures. Wang believes this may reflect aesthetic preferences of the period: he thinks it is unlikely that this difference was present in the original Yongle edition. In any case, while Wang takes the *Siku quanshu* text as the basis for his critical edition, he gives preference to the Jiajing edition illustrations of 1530.[57]

In some respects the *Nongshu* is a conventional farming treatise. Like earlier writers on agriculture, Wang Zhen quotes extensively from earlier and contemporary works, philological, historical, literary and philosophical as well as technical.

56 For a critical analysis of the *Nongshu* editions see Wang Yuhu's study, written in 1966 and included at the beginning of his annotated edition of the treatise (Wang Yuhu, 1981: 3, 5).

57 Wang Yuhu (1981: 9–10, 1–2).

FIGURE 8.4 The emperor ploughs the first furrow of spring in the sacred field outside the capital, 1530 edition of Wang Zhen's *Nongshu*.

Like Lou Shou, he acknowledges the importance of the ritual, symbolic and atmospheric elements of farming. He begins his work conventionally with reflections on *nongben* ("farming is the base"), and on the legendary antecedents of China's agrarian economy. The farming sections of the *Nongshu* are preceded by a depiction of the emperor ploughing the first ritual furrow of spring (Figure 8.4), while the textile sections show the empress and her ladies sacrificing to the Silkworm Deity. A charming disquisition on the value to farmers of the water buffalo includes an allusion to the haunting tones of the herdboy's flute (Figure 8.5). Moreover, Wang like Lou Shou uses poetry, concluding almost every entry in his *Illustrated register* with a few verses by himself or by other poets who had written on farming topics, among them Wang Anshi and Lou Shou.

Wang Zhen's *Nongshu* is, however, an innovation in two key respects. Earlier agronomists such as Jia Sixie, the sixth-century author of the *Qimin yaoshu* (Essential techniques for the common people) or Chen Fu, the Song agronomist whose pioneering short treatise on Jiangnan rice and silk production (also entitled *Nongshu*) appeared in 1149, just after the *Gengzhi tu* was completed, had developed sophisticated technical styles for describing hitherto undescribed farming techniques

FIGURE 8.5 A herdboy plays his flute sitting on the back of a water buffalo, 1530 edition of Wang Zhen's *Nongshu*.

verbally.[58] But although they advocate the use of a wide range of tools and equipment, and explain under which circumstances of season, soil type, weather conditions or growth cycle one type of hoe or harrow rather than another should be used, they appear to take it for granted that their readers will know what all these tools are. Wang Zhen is the first author to highlight the construction of tools and machinery. He is also the first to integrate technical drawings and diagrams with text.

Lou Shou's ballads and genre scenes were paired, certainly, but each made sense independently. Sometimes the pictures were copied without the poems, or one set of poems was replaced by another; the poems themselves also circulated independently of the images, in literary collections or as quotations. But Wang Zhen intended text and image to hold together like a fabric: closely and precisely

58 On the contents and styles of these two works, see, for example, Bray (1984: 55–59) and Bray (2001a, 2001b).

interwoven, strong together though useless apart. 'The images are the warp threads and the words the weft', as Zheng Qiao put it in his essay.[59]

Wang Zhen's pioneering marriage of illustration and explanation constituted a wholly new agronomic rubric: the systematic technical description of field types and of farming equipment. A major section of his treatise, a total of 12 out of 22 chapters, *juan*, in the *Siku quanshu* edition, or 20 out of 36 in the Jiajing edition, pairs illustrations and text to describe varieties of field types, farming implements and equipment, ranging in complexity from water mills to sandals, from harrows to spinning wheels. The *Nongqi tupu* (Illustrated register of agricultural implements) is a superb account of the tools and machines that fed and clothed China. It contains altogether 261 entries, each of which pairs a line drawing of agricultural or textile equipment with text detailing the parts of the tool or machine and their dimensions, how they fit together, and how the tool was used.

What impelled Wang Zhen to develop this new approach to the transmission of knowledge? Wang Zhen was a native of Shandong, and served many years as an official in Anhui and Jiangxi. He was thus acquainted with both the dry-land farming systems of the north and the irrigated rice farming of the south. His purpose in composing the *Illustrated register* was not simply to provide an inventory of the most common farm equipment. Wang's pairing of text and graphics was intended to provide the equivalent of a blueprint, allowing the reader (a fellow official) to reconstruct the equipment (with the help of a skilled carpenter). His hope was that officials would make use of the work to introduce more advanced technology to backward regions, and to disseminate various items of labour-saving or helpful equipment to areas where they were unknown (see Chapter 6). As Zheng Qiao had put it, 'when it comes to assembling implements, without illustrations there is no way to construct them'.[60]

In the opening section of Wang Zhen's work, the *Nongsang tongjue*, this motif of technology transfer occurs again and again. For instance in the chapter on harrowing, *balao*, he quotes a passage from the northern work *Qimin yaoshu*, which advocates using a *ta* (an improvised harrow consisting of a bush or a bunch of branches weighted down with stones) in order to cover the newly sown grain:

> Southerners are not familiar with this implement. Southern practices are different from the north, so they do not know the advantages of using the *ta*. And even in the north, the region is so vast that there are differences there too. Some areas use the *ba* harrow [a harrow with long metal tines] but don't know the *lao* harrow [a harrow consisting of a light hurdle of woven branches], others use the *lao* but don't know the *ba*, and even in the

59 See the quotation that opens this chapter.

60 *Tupu lüe*: 838a.

> north some do not know the *ta*. (For the *ba*, *lao* and *ta*, see also the section *Nongqi pu*.) Now I am including all of them here, so that northern and southern knowledge can be exchanged, and people can use whatever is most suitable.[61]

And indeed the various types of harrow are all meticulously illustrated and described in the corresponding *Nongqi tupu* section.[62]

Here is one labour-saving device that Wang Zhen included in the *Illustrated register* in the hope that it would be widely disseminated:

> The 'weeding oar', *yuntang*, is a new invention from Jiangnan. It is shaped like a wooden patten, a foot or so long and roughly three inches broad, spiked underneath with a dozen or so nails in rows. There is a crossbar on top through which the bamboo handle, five feet or so long, is fixed. At weeding time the farmer stirs together the mud and weeds between the rows of grain so that the weeds are buried in the mud and the field is both cleaned and tilled . . . In the provinces to the east of the Yangzi I have seen farmers weeding their fields by hand, crawling between the crops on their hands and knees with the sun roasting their backs and the mud soaking their limbs, a truly pitiable sight . . . *so I have described the yuntang here in the hope that philanthropists may disseminate its use.*[63]

Alas, perhaps because the *yuntang was* confined at that point to one small region, this passage where Wang clearly states his intention of producing the kind of knowledge that will support technology transfer is also one of the cases where the illustrators of both the Jiajing and the Siku edition, if not perhaps of the lost Yuan edition, failed to match the written description. The tool is, however, correctly illustrated both in the *Nongqi* section of Xu Guangqi's *Nongzheng quanshu* (Complete treatise on agricultural administration) of 1637 and in the corresponding scene in the *Bianmin tuzuan*.[64] Another case where Wang Zhen's illustrator fails to render a piece of rather simple equipment adequately is the cradled scythe used in some northern areas for harvesting wheat and barley; here again, Wang Zhen hoped that this labour-saving device could be introduced into other regions, and used to harvest other cereals.[65] More commonly, however, text and illustration are

61 *WZNS*: 27.
62 *WZNS*: 205–207.
63 *WZNS*: 233, emphasis added.
64 Bray (2007b: Figure 8); *NZQS*: 549; *BMTZ*: 8.
65 *WZNS*: 363–368. According to early twentieth-century observers, notably Amano Motonosuke, the cradled scythe was still in quite common use in the North China plains, but was to be found nowhere else. On the various sources documenting this harvesting device, see Bray (1984: 341).

complementary and sufficiently complete to permit construction and correct use, at least in the case of the simpler implements. Here, for example, is Wang's account of the square flat harrow *ba*:

> The length of the cross-bars should be five feet, the width four inches, the two beams should be five inches or more apart . . . Each is pierced with square holes in which are set wooden teeth six inches or more long. At each end of the cross-bar is a wooden beam about three feet long, curved slightly upwards at the front end and pieced by a wooden peg to which the ox's traces are attached.[66]

In this case the illustration matches perfectly (Figure 8.6). For most of the more complex constructions, including textile equipment and mills, text and illustration complement each other well enough to serve as a reasonable blueprint, an accurate enough guide to allow an experienced carpenter or builder to produce a working

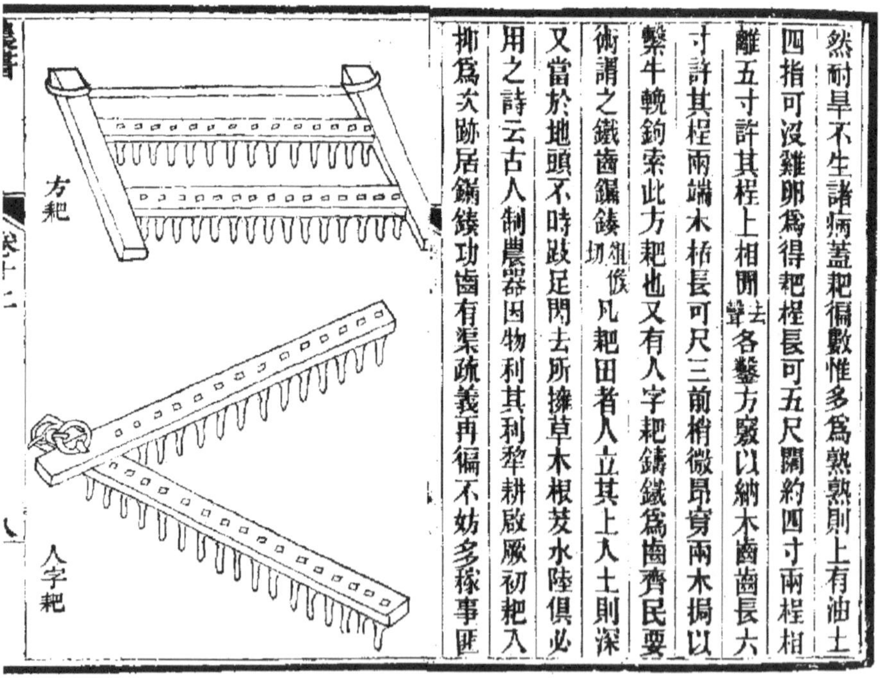

FIGURE 8.6 Square and V-shaped harrows, 1530 edition of Wang Zhen's *Nongshu*.

66 *WZNS*: 205.

version – they are 'schematic representations of a technological principle'.[67] 'This pump involves many joints, but if you have a carpenter, it can easily be made', says Wang Zhen at the end of his description of the chain-pump, *fanche* (Figure 8.7).[68]

Though unprecedented in agronomy, the *Illustrated register* was not a complete innovation. The writing method (naming components, giving their dimensions and describing how they connect) follows the pattern of two short earlier works, Lu Guimeng's *Leisijing* (Classic of the plough) of circa 880, and Qin Guan's *Canshu* (Treatise on sericulture) of circa 1090, both of which Wang quotes.[69] The concept of linking technical text and picture has precedents in Song works such as the

FIGURE 8.7 A chain-pump driven by a water mill, 1530 edition of Wang Zhen's *Nongshu*. This is just one of several water- or animal-powered devices for raising water depicted in the *Nongshu*. The rendering of the linkage between the horizontal mill-shaft and the chain-pump is, to put it politely, schematic; nevertheless, the general principle of harnessing horizontal rotation to vertical drive is conveyed clearly.

67 Kuhn (1977: 144).

68 *WZNS*: 326.

69 Wang Zhen quotes at length both the *Leisijing* description of the plough (*WZNS*: 200–201), and the *Canshu* description of the silk-reeling frame (*WZNS*: 388–390). Lu Guimeng starts the *Leisi jing* by explaining that he had been completely ignorant of the construction of the plough until one day he asked a farmer to tell him what the different parts were called; for a translation and reconstruction, see Bray (1984: 181–182).

Yingzao fashi (Models and methods for building), a guide to the elements and modules of public building compiled by Li Jie at the end of the eleventh century, and the *Kaogongji jie* (Artificer's record explained), a work completed circa 1235 by the scholar Lin Xiyi.[70]

A comparison with the illustrative style of the *Gengzhi tu* highlights the visual choices Wang Zhen has made. The *Gengzhi tu* shows farming tasks as part of a landscape; often the operative parts of plough or harrow are invisible. In the 1530 illustrations to Wang Zhen's *Illustrated register* the components of the implements are all visible. Often the implement or machine stands alone, and where persons or animals are included it is to demonstrate its operation (see Figure 3.1). Only in the case of field types is the surrounding landscape shown, with small figures hurrying along the bunds and cottages tucked behind a clump of bamboos (see Figure 2.3). The illustrations of silk-processing and weaving, however, not only show the women working the machines but also, in several cases, add some non-functional scenery; the similarity of the representations to those in the early versions of the *Gengzhi tu* suggests that where there were visual precedents, Wang Zhen's woodblock carvers resorted to their use.[71] This is also true for instance of the scene of storing grain, where the Jiajing illustration of the scene is an almost exact copy of the *Gengzhi tu* version.[72]

As already mentioned, even though his illustrations of equipment owe almost nothing to the pictures in the *Gengzhi tu*, Wang Zhen does quote from Lou Shou's poems; for example, Lou's poem on flailing the grain, translated above, which Wang Zhen includes after his technical description of the flail. Wang Zhen says that the flail is used all through the south but seldom in the north, and he advocates its adoption as an easy way of husking any grain. Describing 'its construction' (*qi zhi*), he says:

> Four thin wooden sticks are used, bound together with rawhide; the bundle should be about 3 feet long and four inches wide. Sometimes a single heavy stick is used. Both kinds have a long wooden handle, the top of which is pierced to take a spindle, allowing the flail to spin round as it is lifted to strike the grain.[73]

70 Lin attempts to interpret the notoriously cryptic *Kaogongji* (Artificer's record), a section from the *Zhouli* (Rituals of the state of Zhou) that may date back to the Warring States period. Lin Xiyi's illustration of his reconstruction of the Zhou plough, *leisi*, is shown in Bray (1984: 145–146).

71 For example, the depictions of silk reels and looms, *WZNS*: 404–407. The illustrations of cotton reeling equipment (*WZNS*: 417–419), which was relatively new to China when Wang Zhen composed his work, appear to have given the artist far more trouble: they are angular and awkward, unlike the fluid lines and easy postures of the sericultural pictures.

72 See Bray (2007b: Figure 12), comparing *WZNS*: 290 and Franke (1913: Figures I, 21).

73 *WZNS*: 253–254. Though Wang's explanation is very clear, his illustration is not nearly as useful as Lou Shou's.

How did Wang Zhen devise his particular method of visually and verbally inscribing technical knowledge? Unfortunately, he himself does not tell us. Jia Sixie, the sixth-century author of the *Qimin yaoshu*, tells us in his preface that he consulted other farmers; the seventeenth-century agronomist Xu Guangqi (discussed below) questioned peasants relentlessly about technical details, taking copious notes all the while; both were enthusiastic experimenters.[74] As to Wang Zhen's procedures, we may infer them both from his treatise and from the preface that he asked his friend and fellow prefect Dai Biaoyuan to write for the *Nongshu*. Dai writes that he first heard about Wang Zhen when he was prefect of Jingde in Anhui:

> I asked him about his methods. Each year he instructed the common people in a certain number of elements of farming and textile production. He thus transmitted to them all the methods, *fang*, of growing and harvesting fibre crops and grain. He also made drawings, *tuhua*, of all the varieties of hoes, drills, harrows, and other miscellaneous implements, and had the common people make them, *shi min wei zhi*. At first the common people said: 'But this is our trade, we have done this for generations. Why should we need instruction?' The administrators in other prefectures all ridiculed him, considering this was really not part of his job: a good official, a wise administrator, would know all this without asking, and if he did ask, he would not claim it as an achievement. But Boshan [Wang Zhen] continued for three years, and before he was transferred to his next post the common people of Jingde had come to rely on him and were singing his praises. Boshan not only instructed them in the methods, *fang*, and implements, *qi*, of farming, he also was able to do so in such an unintrusive and peaceable way that they improved daily and with docile hearts.[75]

Following the Confucian ideal of government, which presumed that knowledge passed down from the government to the people, Dai suggests that Wang Zhen taught the local peasants much, and they taught him nothing. Wise rulers of yore had taught the people to build houses, to make clothes, to cook food, to grow millet; naturally, then, wise magistrates of the Yuan dynasty taught the local people how to make a chain-pump or a flail rather than the other way around. But how did Wang Zhen come by this technical knowledge? It seems likely that Dai Biaoyuan is conflating a process of give and take of knowledge into a one-way flow. Presumably, wherever Wang Zhen was posted he made a point not only of sketching the local tools and equipment and noting their construction, but also of asking the

74 *QMYS*: 5; on Xu Guangqi's techniques for interviewing farmers, see Shi Shenghan's introduction to his edition of the *Nongzheng quanshu* (Shi, 1979: 4).

75 *Wang Boshan nongshu xu* (Preface to Wang Boshan's Agricultural treatise), appended in Wang Yuhu's edition to the main text, *WZNS*: 445.

local craftsmen to take the implements apart and put them together again so that he could be sure he understood well enough to make, in Dieter Kuhn's terms, 'a schematic representation of a technological principle'. Perhaps he also tried to persuade people to try building and using equipment they were unfamiliar with, or to make improved versions of known tools (as Dai's account suggests), but if so, Wang himself does not tell us of his successes, only of his hopes. There is no doubt that new methods and new equipment did spread from one region to another through the late imperial period, but it is not clear that Wang Zhen's *Illustrated register*, brilliant innovation though it was, actually played any direct role in promoting specific cases of technological transfer.[76]

Yet Wang Zhen's paired texts and images, like Lou Shou's, were widely diffused. Apart from the many later editions of the *Nongshu* itself, the whole of Wang Zhen's *Illustrated register* (the Jiajing version) was included in two important later works on agronomy, the *Nongzheng quanshu* (Complete treatise on agricultural administration) of 1639, by the late Ming statesman-polymath Xu Guangqi, and the imperial compilation *Shoushi tongkao* (Compendium of works and days) of 1742.[77] Numerous entries from the *Illustrated register* were incorporated into Wang Qi's influential *Sancai tuhui* (Illustrated encyclopedia of Heaven, Earth and Man) of 1609, and the illustrations thus entered common currency as emblematic representations of well-known everyday objects such as ploughs, hoes or looms. The *Sancai tuhui* seems to have included few or no original illustrations, instead incorporating images drawn from quite a wide range of illustrated technical works. As far as farming and textiles are concerned, the *Sancai tuhui* seems to have taken the *Nongshu* as its only source. Though it does include one or two scenes or motifs found originally in the *Gengzhi tu*, they are the versions found in the Jiajing edition of the *Nongshu*.[78] The *Nongshu* never equalled the *Gengzhi tu* in its impact on Korean or Japanese farm treatises or rural iconography, yet a number of the *Nongshu* illustrations did find their way to Japan via the *Sancai tuhui*. The *Shinzoku kibun* (Recorded accounts of Qing customs), for instance, an account of everyday life, customs and beliefs compiled on the order of the Governor of Nagasaki in 1799, drew heavily on the *Sancai tuhui* and included a number of its illustrations of farming equipment and textile production.[79]

76 The possible vectors of technical innovation are discussed at greater length in Chapter 6.

77 As mentioned earlier, the *Gengzhi tu* was also included in the *Shoushi tongkao*.

78 *Nongshu* depictions of silkworm sheds and of the imperial ceremonies concerning sericulture, and of granaries, stables and farm buildings, are included in the *Sancai tuhui* section on "Palaces and buildings" (vol 3: 1017–1025); pictures of textile equipment, irrigation devices and mills, and agricultural tools are included in the section on "Useful devices" (vol 3: 1257–1233). The scene of storing grain (vol 3: 1020) is one example of an indirect borrowing, through the *Nongshu*, from the *Gengzhi tu*.

79 See *SZKB*: 197–213; on how this Japanese work was produced, see Chapter 4.

Taking things further?[80]

Both the *Gengzhi tu* and the *Illustrated register* were creative innovations in the verbal-visual depiction of technical activities. But were these innovative impulses carried any further?

Let us first consider the case of Xu Guangqi's *Nongzheng quanshu* (Complete treatise on agricultural administration) of 1639. As discussed in Chapters 6 and 7, Xu Guangqi (1562–1633) was an eminent statesman and a critical and creative polymath. Most of Xu's voluminous writings on agriculture were incorporated into the *Nongzheng quanshu*, a compendium posthumously edited and published by a group of Xu's young admirers, who shared his view that agricultural reform and improvement were of the first importance for restoring the ailing Ming regime to health and strength.

Xu's research on agriculture was innovative in several respects: he used historical records to analyse natural patterns, for instance, in his work on how to predict locust swarms; he experimented with the introduction of plants from one climatic zone to another, as in the case of the sweet potato, which he successfully persuaded farmers in the Yangzi region to adopt. Unlike many agronomic authors who were content simply to quote from earlier authorities, Xu Guangqi was nothing if not critical. He spent considerable time and effort himself experimenting and testing farming methods, documenting regional practices and discussing them with farmers. He was a hands-on man who felt entitled to remark with some asperity that Wang Zhen was a better poet than he was a farmer.[81] Nevertheless, when it comes to the description of equipment, Xu includes Wang's *Illustrated register* almost in its entirety in his own treatise, although it is re-ordered and broken up into separate sections.[82]

The changes made by Xu to Wang's original are as follows: he drops all the poems, adds later materials on the same topic, and appends his own comments. In the section on various types of water-raising equipment, for example, Xu adds notes on which devices are most suitable for particular speeds or depths of water flow.[83]

80 For reasons of space I have had to omit from this section another interestingly innovative work, Fang Guancheng's *Mianhua tu* (Pictures of cotton) of 1765. I discuss Fang's graphic and verbal techniques, and the scope and limitations of his graphic and textual innovations, in Bray (2007b: 548–549).

81 *NZQS*: 123.

82 The editors of the *Nongzheng quanshu* saw it as a primary duty to remain faithful to Xu's intentions, and later scholars have shown that they made almost no substantive changes to the information and opinions expressed in Xu's manuscripts. The work by Chinese scholars on this issue is summarized in Bray and Métailié (2001).

83 See, for example, Xu's detailed remarks on the suitable conditions for installing a noria, *tongche*, rather than some kind of chain-pump for raising water, *NZQS*: 425.

The illustrations are closely based on those in the Jiajing edition of Wang Zhen's *Nongshu*, though more roughly drawn, so that sometimes the technical features are less easy to understand. Some of Wang Zhen's originals are omitted, and some have undergone minor but intriguing modifications, as in the picture of the chain-pump, which shows one of the two men reading a book as he pedals (Figure 8.8).[84] Some new illustrations are added in Xu's treatise, notably the diagrams of Archimedes' screw and other Western pumping devices to which Xu had been introduced by his Jesuit colleagues, and the drawings of plants in the long final

FIGURE 8.8 Chain-pump with worker reading, 1639 edition of *Nongzheng quanshu*.

84 For example, Wang Zhen's illustration of the silk-reeling machine is referred to in Xu's text but not included, *NZQS*: 861, 878. In the illustration of the chain-pump, the axle that drives the pump is severely distorted, though the artist lovingly details one man's book and the other's parasol, *NZQS*: 423. The relative crudeness of the illustrations is no doubt due to the haste with which the editors pushed through publication of the work; modern scholars have remarked that the carving of the blocks for both text and illustrations was far from perfect (Liang, 1963: 105–106).

section on famine foods.[85] There had been some significant technical changes and improvements between Wang Zhen's lifetime and Xu Guangqi's, for example to cotton spinning and silk-reeling equipment, and Xu's text mentions a number of such technical developments, but unfortunately none are reflected in the illustrations.[86]

Let us now consider Song Yingxing's wonderful book on crafts and manufactures, the *Tiangong kaiwu* (The exploitation of the works of nature), a work renowned for its originality. It is probable that the illustrations for the original edition of 1637 were produced in several workshops, or at any rate by more than one woodblock artist.[87] Most of the pictures of farming tasks, irrigation equipment, food processing and textile production were original in conception, and quite different from all the well-known earlier representations in the tradition of the *Gengzhi tu* or the *Nongshu*.[88] Nor does it seem likely that most of these quirky drawings, full of vitality, were copied from other existing pictures. A case in point is the earliest known depiction of the elaborate and expensive machine that Kuhn calls the "complete silk-reeling frame"(Figure 8.9), which must just have been coming into common use at the time when the *Tiangong kaiwu* was written.[89] Whatever the accuracy of their technical details, or the degree to which the drawings corresponded to Song's text,[90] this originality in itself is quite astonishing for the period. However, a few of the 1637 drawings are more typical of the common illustrative practices of the period in that they seem to be crude and hasty copies of *Nongshu* pictures.[91] Could they be the product of a different workshop from those which generated entirely new images of technical processes? It is also notable that when Chen Menglei selected excerpts from the *Tiangong kaiwu* to include in his *Gujin tushu jicheng* (Illustrated compendium of texts and pictures, ancient and modern) of 1725, he

85 *Juan* 19 and 20 of the *Nongzheng quanshu* are devoted to *Taixi shuifa* (Western irrigation methods). *Juan* 43-60 describe and illustrate edible plants that can be used as famine foods; this whole section is based on the earlier treatise *Jiuhuang bencao* (Inventory of plants for use in famines), by the Ming prince Zhu Xiao, completed in 1406 and reprinted several times in the later Ming.

86 As the modern commentator Shi Shenghan forlornly remarks, e.g. *NZQS*: 878.

87 See Golas (2007).

88 The *Tiangong kaiwu* includes illustrations of numerous occupations and techniques never before illustrated, such as pearl diving, grinding sugar cane or the casting of bells.

89 This machine became standard in urban workshops between about 1600 and 1800. Its sophisticated design automated the adjustment of rollers, guiding-eyes, etc., so that the operator had only to concentrate on the reels, and was thus able to reel between 600 and 1,000 grams of thread a day. It greatly increased efficiency and facilitated mass production (Kuhn, 1988: 390–400).

90 See Golas (2007) and Wagner (2007).

91 By the late Ming, there was a craze for illustrations, but the repertory that artists or publishers drew on was surprisingly small (Hegel, 1998). Illustrations circulated between genres: a virtuous woman serving medicine to her mother-in-law could easily be redrawn to show a lovelorn girl reading a letter from her suitor (Carlitz, 1991; Raphals, 1998). Prefacing the *Bianmin tuzuan* with an adapted *Gengzhi tu* was both a tribute to the virtues of rural activity, and a way of attracting purchasers.

FIGURE 8.9 Complete silk-reeling frame, 1637 edition of *Tiangong kaiwu*. Although the accompanying text is technically precise, the illustration is impressionistic and a chair has somehow become entangled in the worker's legs.

added a set of new and stylistically more elaborate woodblock illustrations. Here we find that most of the farming, irrigation and textile scenes are not original but are close copies from either the *Gengzhi tu* or the *Illustrated register*.[92]

The illustrations of irrigation devices in the 1637 edition of *Tiangong kaiwu* are five in number, showing the noria, *tongche*, the ox-powered chain-pump, *niuche*, and the two-man-powered chain-pump, *tache*, a chain-pump operated by hand by a single person, *bache*, and the well-sweep, *jiegao*.[93] Song's text on irrigation is very brief, just a few columns; he names the five types of equipment at the beginning,

92 See, for example, the 1936 Commercial Press edition from the *Guoxue jiben congshu jianbian* series (Song Yingxing, 1936), and the English translation (Sun and Sun, 1966), which include both sets of illustrations.

93 The Qing version also has illustrations of an elevated noria, a tank, a coffer-dam and a chain-pump driven by a water mill (Sun and Sun, 1966: 15–18), none of which are actually described in the *Tiangong kaiwu* text although all are both illustrated and described in Wang Zhen's *Nongshu*.

states that each is illustrated, then concentrates in particular on the inputs of labour that each type of equipment requires to water a given acreage.[94] Of the set of five Ming woodblocks only one is recognizably a copy, namely the two-man chain-pump, which is a very crude rendering of the *Nongshu* drawing; though the axle-mechanism is clearer than in the *Nongzheng quanshu* rendering, the chain itself is so roughly drawn as to be meaningless. The hand-operated pump, *bache*, on the other hand, is presented here for the first time, and drawing and text taken together make it perfectly plain how this ingenious and convenient device worked (Figure 8.10). It is a portable device, says Song, ideal for ponds or channels too shallow for a chain-pump. The illustration shows a wheel, mounted vertically on a stand, with a cranked handle attached to either side of the hub; the text tells us that a single person works the pump, pulling each crank back in rapid alternation. Song does not describe the *bache* as being new, but it was the kind of machine that would

FIGURE 8.10 Hand chain-pump, *bache*, 1637 edition of *Tiangong kaiwu*.

94 Song Yingxing (1936: 4–5); Yabuuchi (1969: 26–30).

serve well where farms and fields were small, and water and labour resources scarce, as they increasingly were in the fertile lower Yangzi by the end of the Ming. [95]

Another interesting original twist in the Ming illustrations of the *Tiangong kaiwu* occurs in the sections on textile production. Here the woodblock artists show a significant shift from the traditional gender division of labour, not mentioned specifically in Song's text but prominent in other written sources of the period. Several pictures show men at work where earlier illustrations had always showed women (see Figure 3.2). Though the *Tiangong kaiwu* still shows women raising silkworms, it shows men reeling and twisting yarn, mounting the loom and weaving the cloth – an accurate reflection of the silk industry of the late Ming, which was now largely based in urban workshops that depended on male wage-labourers. This change was certainly not reflected in the woodblock illustrations of textile production in the *Nongzheng quanshu*, published just two years later, also in Jiangnan.[96]

Concluding remarks

The popularization of woodblock printing in the Song opened up new possibilities for inscribing and disseminating technical information, most notably new opportunities for illustration. The unique characteristics and possibilities of woodblock printing played a key role in the development and dissemination of agricultural knowledge in China, and woodblock illustration was among the pedagogical tools developed in these early centuries of print culture. Wang Zhen clearly designed his treatise, and in particular the *Illustrated register*, as a work to be printed, with text and image inseparably paired. Wang Zhen's marriage of text and image was designed to convey material details simply and with clarity, and his descriptions were intended as blueprints, documents that could be used to construct an unfamiliar piece of machinery or equipment. To that extent, Wang Zhen was engaged in what today we would call technical drawing.

Lou Shou's poems provided rather little in the way of hard technical information, and his paintings were scenes not diagrams, though they depicted the advanced methods of Jiangnan in considerable and accurate detail. The *Gengzhi tu* was considered by the ruling class of the time to have considerable pedagogic value, and it was therefore very soon translated into woodblock versions for convenient

95 In fact the *Illustrated register* includes an entry on a similar device called the *guache* ("slicing pump", so called, says Wang Zhen, because the vanes of the wheel slice up the water), which closely resembles the *bache* except that it is mechanically less efficient, being worked with a single crank on one side of the axle (*WZNS*: 336).

96 Yabuuchi (1969: 64–71); *NZQS*: 946-949. On changes in the gender division of labour in the textile industry, see Chapter 3.

dissemination. From the beginning, its pedagogical content was seen as two-fold. First the combined poems and images were thought of as technical descriptions that could circulate images of sophisticated Jiangnan rice farming and silk-producing methods to less advanced regions. But second, they were also valued because they so vividly crystallized iconic scenes of an idealized rural landscape, and of the social and political order that such a landscape was supposed to sustain.

In early-modern Europe, technical drawing has been identified as a catalyst of invention, encouraging craftsmen and engineers to think in new ways and to imagine new devices. Yet in late imperial China one could argue that illustration often played a conservative or even an archaicizing role. Even Wang Zhen recorded only what was already in existence; he did not develop or invent new machines in the workshop, nor did he, or the artists hired to execute the woodblocks for his work, doodle ingenious new devices on paper. Still more noteworthy is the fact that although the illustrations to Wang Zhen's *Illustrated register* were widely known and frequently reproduced in late Ming and Qing China, most later agronomists were content simply to have their woodblock artists copy the entries rather than to amend or build upon them. Xu Guangqi added many observations and qualifications to Wang's entries, but all were expressed in verbal, not visual, form. It is true that mechanical innovations in late imperial China were rather few in number, but there were some. However, the only ones for which I have been able to find illustrations are the new hand-pump and spinning frame shown in the *Tiangong kaiwu*. Furthermore, we see almost no attempts in later agricultural works to develop Wang Zhen's illustrative techniques into more sophisticated forms; instead most authors found it sufficient to translate new or improved technical procedures into words.

The *Tiangong kaiwu* is quite exceptional in this respect. In many cases Song Yingxing's woodblock artists produced completely new visual renderings of equipment rather than simply copying earlier versions. They also produced many original additions to the existing repertory of technical activities. Otherwise, however, we find almost no attempts to expand or to improve the repertory or methods of technical illustration. Instead, the practice of simply copying earlier pictures usually prevailed, and in the process, many technical depictions became less rather than more accurate.

This is of little importance if the depictions are valued primarily as symbols. Given their prominent moral content and their self-consciously artistic form of expression, it is easy to understand how the images and poems of the *Gengzhi tu* quickly acquired iconic status as timeless representations of an idealized order. When Lou Shou produced the *Gengzhi tu,* the techniques he portrayed were among the most advanced in the country. But 600 years later, the Qing emperors valued them not for any novel technical content but rather for their moral force and purity.

The *Gengzhi tu* images of rural life were not originally designed in the medium of print, yet printing and analogous forms of reproduction played a key role in enhancing the reach and power of these icons. The boundaries between different inscriptive media, and different genres of communication, were very fluid, and although woodblock printing played an important role in early-modern China as a medium for the mass circulation of new information, it also served to consolidate traditions. The *Gengzhi tu* began as paintings on silk, paired with calligraphed poems. In other words, it was executed in the genre of the album, usually a rather intimate medium of communication between educated friends or colleagues. However, the technique of transcribing such artistic works into stone carvings, from which rubbings could be made, was an ancient one, and by the Song woodblock printing was also routinely used to transcribe paintings and calligraphy for a wider audience. In its printed forms the *Gengzhi tu* quickly achieved wide circulation, and its scenes of farming and weaving soon became part of a visual repertory that was drawn upon by artists at every level of society when they wished to show typical scenes of rural life. It is interesting that several woodblock graphics from Wang Zhen's much more utilitarian *Nongshu* acquired similar iconic status as representations not of technology but of lifestyle, through their incorporation into popular illustrated encyclopedias such as the *Sancai tuhui*.

The ruling classes of late imperial China never abandoned their efforts to "promote agriculture". Yet one of their favourite vehicles for such campaigns, right into the nineteenth century, was Lou Shou's *Gengzhi tu*, with a new set of poems added by the reigning emperor, and perhaps a dash of Western perspective or a couple of new scenes. Wang Zhen's *Illustrated register* offered much more innovative visual possibilities, but they were taken no further. When it came to practical, technical matters such as farming, it seems that those educated Chinese who recorded changes or improvements were satisfied with the power of words to convey material processes, and felt no need or desire to resort to graphics.

REFERENCES

Primary sources

BMTZ Bianmin tuzuan (Collection of pictures for the convenience of the people), 1st ed. 1502, attr. Guang Fan. Refs are to modern annotated edn. with facsimile of 1593 woodblock illustrations, Shi Shenghan and Kang Chengyi (1982 [1959]).

BNS Bu nongshu (Supplemented treatise on agriculture), Zhang Lüxiang, 1658. Incorporating a treatise on agriculture by an anonymous Mr Shen (*Shenshi Nongshu*), probably written just before 1640. Modern annotated edn. Chen Hengli and Wang Dacan (1983).

CFNS Nongshu (Agricultural treatise), Chen Fu, preface 1149. Repr. Beijing: Zhonghua Editions, 1956.

Fan Shengzhi shu (Book of Fan Shengzhi), c. 160 CE. Modern annotated edn. Shi Shenghan (1959).

Gengzhi tu (Pictures of tilling and weaving), Lou Shou, presented to the emperor in 1145. Reproduced in Franke (1913), Pelliot (1913) and Hammers (2011).

Leisi jing (Classic of the plough), Lu Guimeng, c. 880. Repr. Beijing: Zhonghua Press, 1956.

Lu Ban jing (Canon of Lu Ban, patron saint of carpenters, or Carpenters' canon), various dates. See Ruitenbeek (1993) for dating and complete text.

LSCQ Lüshi chunqiu (Spring and autumn annals of Master Lü [Buwei]), c. 239 BCE. Modern annotated edn. Xia Weiying (1956).

LZZN Linzheng zhinan yi'an (Medical records as a guide to diagnosis), Ye Gui (Tianshi), 1776. Repr. Taipei: Chongde Bookshop, 1984.

Mianhua tu (Pictures of cotton), Fang Guancheng, presented to the Qianlong emperor in 1765; pr. as *Shouyi guangxun*.

Nōgyō zensho (Collected writings on agriculture), Miyazaki Yasusada, preface 1697.

NSJ Nongsang jing (Canon of farming and sericulture), Pu Songling, preface 1705. Refs are to modern edn. Li Changnian (1982).

NSJY Nongsang jiyao (Compiled essentials of agriculture and sericulture), preface by Wang Pan, probable editor Meng Qi, 1273.

NZQS Nongzheng quanshu (Complete treatise on agricultural administration), Xu Guangqi, ed. Chen Zilong *et al.* (1639). Refs are to modern annotated edn. Shi Shenghan (1979).

QMYS Qimin yaoshu (Essential techniques for the common people), Jia Sixie, c. 635. Refs are to modern annotated edn. Miao and Miao (1982).

Sancai tuhui (Illustrated encyclopedia of Heaven, Earth and Man), Wang Qi, pref. 1609; facsimile edn., Taipei: Chengwen Press, 1970, 10 vols.

Shangyou tang yi'an (Medical case histories from the Hall of Honored Friendship), Fang Lüe, 1846. Repr. Shanghai: Rare Books Press, 1983.

Shoushi tongkao (Compendium of works and days), comp. E'ertai, presented to the throne 1742; mod. edn. Beijing: Nongye Press, 1963.

Shouyi guangxun (Expanded instructions on producing clothing), printed edn. of *Mianhua tu*, with additional poems and comments, produced under the direction of Dong Gao and presented to the Jiaqing emperor in 1808; facsimile edn. *Zhongguo gudai banhua congkan*, vol. 4, Beijing: Zhonghua Press, 1960.

Simin yueling (Monthly ordinances for the four classes of people), Cui Shi, c. 160 CE. Annotated modern edn. Shi (1965).

Sishi zuanyao (Compendium of essentials for the four seasons), Han E, possibly late ninth century; modern annotated edn. Miao Qiyu (1981b).

SZKB Shinzoku kibun (Recorded accounts of Qing customs), commissioned by Nakagawa Tadahide, Nagasaki, 1799; facsimile edn. Taipei: Tali Press, 1983.

TGKW Tiangong kaiwu (The exploitation of the works of nature), Song Yingxing, 1st edn. 1637; editions referred to here are Song Yingxing (1936), Yabuuchi (1969), Pan Jixing (1989), and the English tr. by Sun and Sun (1966).

Tupu lüe (A brief account of illustrated registers), c. 1150, *Tongzhi* (Comprehensive studies) *j*. 71/837–840, by Zheng Qiao; Taipei: Xin Xing Shuju, no date, 10 vols. (Collection: *Guoxue jiben congshu*).

WZNS Nongshu (Agricultural treatise), Wang Zhen, preface 1313. Refs are to modern annotated edn. Wang Yuhu (1981).

XQOJ Xianqing ouji (Casual expressions of idle feeling), Li Yu, 1671. Facsimile edn. Taipei: Guangwen Press, 1977.

YLLS Yilue liushu, nüke zhiyan (Medical compendium in six books, Experiences in gynecology), Xu Dachun, 18th century. Refs are to *Xu Dachun yishu quanji* (Complete medical writings of Xu Dachun), ed. Beijingshi Weisheng Ganbu Jinxiuyuan Zhongyibu, 2 vols, Beijing: Renmin Weisheng Chubanshe, 1988.

Other references

Abu-Lughod, Lila (1986) *Veiled Sentiments: Honor and Poetry in a Bedouin Society*, Berkeley, CA: University of California Press.

Adas, Michael (1989) *Machines as the Measure of Men: Science, Technology, and Ideologies of Western Dominance*, Ithaca, NY: Cornell University Press.

Akrich, Madeleine (1993) 'A gazogene in Costa Rica: an experiment in techno-sociology', in Pierre Lemonnier (ed.), *Elements for an Anthropology of Technology*, Ann Arbor, MI: Museum of Anthropology, University of Michigan, pp. 289–337.

Amano, Motonosuke (1979) *Chūgoku nōgyōshi kenkyū* (Researches into Chinese agricultural history), 2nd expanded edn. (1st edn. 1962), Tokyo: Ryukei Press.

Ardener, Shirley (ed.) (1993) *Women and Space: Ground Rules and Social Maps*, 2nd edn., Oxford: Berg.

Arrighi, Giovanni (2007) *Adam Smith in Beijing: Lineages of the Twenty-First Century*, London: Verso.

Arrighi, Giovanni, Takeshi Hamashita and Mark Selden (eds) (2003) *The Resurgence of East Asia: 500, 150 and 50 Year Perspectives*, London and New York: Routledge.

Aunger, Robert (2010) 'What's special about human technology?', *Cambridge Journal of Economics* 34: 115–123.

Bachelard, Gaston (1957) *La poétique de l'espace*, Paris: Presses Universitaires de France.

Bankoff, Greg and Peter Boomgaard (eds) (2007a) *A History of Natural Resources in Asia: The Wealth of Nature*, New York: Palgrave Macmillan.

—— (2007b) 'Introduction: natural resources and the shape of Asian history, 1500–2000', in Greg Bankoff and Peter Boomgaard (eds), *A History of Natural Resources in Asia: The Wealth of Nature*, New York: Palgrave Macmillan, pp. 1–17.

Barbieri-Low, Anthony (2007) *Artisans in Early Imperial China*, Seattle, WA: University of Washington Press.

Basalla, George (1988) *The Evolution of Technology*, Cambridge: Cambridge University Press.

Beattie, Hilary (1979) *Land and Lineage in China: A Study of T'ung-ch'eng County, Anhwei, in the Ming and Ch'ing Dynasties*, Cambridge: Cambridge University Press.

Bennett, Steven J. (1978) 'Patterns of the sky and earth: a Chinese science of applied cosmology', *Chinese Science* 3: 1–26.

Berger, Günther, Georges Métailié and Takeshi Watabe (1996) 'Une chinoiserie insolite: étude d'un papier peint chinois', *Arts Asiatiques* 51: 96–116.

Berliner, Nancy (ed.) (1996) *Beyond the Screen: Chinese Furniture of the 16th and 17th Centuries*, Boston, MA: Museum of Fine Arts.

Bijker, Wiebe E. (2007) 'American and Dutch coastal engineering: differences in risk conception and differences in technological cultures', *Social Studies of Science* 37, 1: 143–152.

—— (2010) 'How is technology made? That is the question!', *Cambridge Journal of Economics* 34: 63–76.

Blake, C. Fred (1994) 'Foot-binding in neo-Confucian China and the appropriation of female labor', *Signs* 19, 3: 676–712.

Boomgaard, Peter (2007) 'From riches to rags? Rice production and trade in Asia, particularly Indonesia, 1500-1950', in Greg Bankoff and Peter Boomgard (eds), *A History of Natural Resources in Asia: The Wealth of Nature*, New York: Palgrave Macmillan, pp. 185–203.

Bourdieu, Pierre (1973) 'The Berber house', in Mary Douglas (ed.), *Rules and Meanings*, Harmondsworth: Penguin, pp. 98–110.

—— (1976) (tr. Richard Nice) *Outline of a Theory of Practice*, Cambridge: Cambridge University Press.

—— (1984) (tr. Richard Nice) *Distinction: A Social Critique of the Judgement of Taste*, London: Routledge.

—— (1990) (tr. Richard Nice) *The Logic of Practice*, Palo Alto, CA: Stanford University Press.

Bray, Francesca (1984) *Science and Civilisation in China*, Volume VI Part 2, *Agricultures*, Cambridge: Cambridge University Press.

—— (1986) *The Rice Economies: Technology and Development in Asian Societies*, Oxford: Basil Blackwell (repr. 1994, Berkeley, CA: University of California Press).

—— (1995) 'A deathly disorder: understanding women's health in late imperial China', in Donald Bates (ed.), *Knowledge and the Scholarly Medical Traditions*, Cambridge: Cambridge University Press, pp. 235–250.

—— (1997) *Technology and Gender: Fabrics of Power in Late Imperial China*, Berkeley, CA: University of California Press.

—— (1998) 'Technics and civilization in Late Imperial China: an essay in the cultural history of technology', *Osiris* 13 (Winter 1998/99): 11–33.

—— (1999) 'Towards a critical history of non-Western technology', in Timothy Brook and Gregory Blue (eds), *China and Historical Capitalism: Genealogies of Sinological Knowledge*, Cambridge: Cambridge University Press: 158–209.

—— (2001a) 'Technique essenziale per il populo' (Essential techniques for the common people), in Karine Chemla, Francesca Bray, Daiwie Fu, Yilong Huang and Georges Métailié (eds), *La storia della scienza in Cina*, in Sandro Petruccioli (ed.), *Enciclopedia internazionale della historia della scienza*, Rome, Istituto Treccani, Volume II: 1–608, pp. 208–219.

—— (2001b) 'L'agricoltora', in Karine Chemla, Francesca Bray, Daiwie Fu, Yilong Huang and Georges Métailié (eds), *La storia della scienza in Cina,* in Sandro Petruccioli (ed.), *Enciclopedia internazionale della historia della scienza*, Rome, Istituto Treccani, Volume II: 1–608, pp. 501–511.

—— (2003) 'Genetically modified foods: shared risk and global action', in Barbara Herr Harthorn and Laury Oaks (eds), *Revising Risk: Health Inequalities and Shifting Perceptions of Danger and Blame*, Westport, CT: Praeger, pp. 185–207.

—— (2005) 'The inner quarters: oppression or freedom?', in Ronald G. Knapp and Kai-Yin Lo (eds), *House Home Family: Living and Being Chinese*, Honolulu, HI: University of Hawaii Press, pp. 259–280.

—— (2007a) 'Introduction: the powers of *tu*', in Francesca Bray,Vera Dorofeeva-Lichtmann and Georges Métailié (eds), *Graphics and Text in the Production of Technical Knowledge in China*, Leiden: Brill, pp. 1–78.

—— (2007b) 'Agricultural illustrations: blueprint or icon?', in Francesca Bray,Vera Dorofeeva-Lichtmann and Georges Métailié (eds), *Graphics and Text in the Production of Technical Knowledge in China*, Leiden: Brill, pp. 526–574.

—— (2007c) 'Gender and technology', *Annual Review of Anthropology* 36: 37–53.

—— (2008a) 'The technics of the America home', in Sandra Montón-Subías and Margarita Sánchez-Romero (eds), *Engendering Social Dynamics: The Archaeology of Maintenance Activities*, Oxford: BAR International Series 1862 (Hadrian Books), pp. 25–31.

—— (ed.) (2008b) *East Asian Science, Technology and Society: An International Journal*, special issue, *Constructing Intimacy: Technology, Family and Gender in East Asia*, 2, 2.

—— (2010) 'How blind *is* love? Simon Winchester's *The Man Who Loved China*', *Technology and Culture* 51, 3: 327–333.

—— (2013) 'Tools for virtuous action: technology, skills and ordinary ethics', in Charles Stafford (ed.) (2013), *Ordinary Ethics in China*, London: Berg, pp. 175–193.

—— (forthcoming) 'Agriculture', in Albert Dien and Keith Knapp (eds), *The Cambridge History of China, The Six Dynasties, 220–581* (forthcoming).

Bray, Francesca, Peter Coclanis, Edda Fields-Black and Dagmar Schäfer (eds) (2014, in press) *Rice: Global Networks and New Histories*, New York: Cambridge University Press.

Bray, Francesca, Vera Dorofeeva-Lichtmann and Georges Métailié (eds) (2007) *Graphics and Text in the Production of Technical Knowledge in China*, Leiden: Brill.

Bray, Francesca and Georges Métailié (2001) 'Who was the author of the *Nongzheng quanshu*?', in Catherine Jami, Peter Engelfriet and Gregory Blue (eds), *Statecraft and Intellectual Renewal in Late Ming China: The Cross-Cultural Synthesis of Xu Guangqi (1562–1633)*, Leiden: Brill, pp. 322–359.

Brenner, Robert and Chris Isett (2002) 'England's divergence from China's Yangzi Delta: property relations, micro-economics, and patterns of development', *Journal of Asian Studies* 61, 2: 609–662.

Brokaw, Cynthia J. and Kai-wing Chow (2005) *Printing and Book Culture in Late Imperial China*, Berkeley, CA: University of California Press.

Brook, Timothy (1995) 'Weber, Mencius, and the history of Chinese capitalism', *Asian Perspective* 19, 1: 79–97.

—— (1998) *The Confusions of Pleasure: Commerce and Culture in Ming China*, Berkeley, CA: University of California Press.

—— (1999) 'Capitalism and the writing of modern history in China', in Timothy Brook and Gregory Blue (eds), *China and Historical Capitalism: Genealogies of Sinological Knowledge*, Cambridge: Cambridge University Press, pp. 110–157.

—— (2001) 'Xu Guangqi in his context: the world of the Shanghai gentry', in Catherine Jami, Peter Engelfriet and Gregory Blue (eds), *Statecraft and Intellectual Renewal in Late Ming China: the Cross-Cultural Synthesis of Xu Guangqi (1562–1633)*, Leiden: Brill, pp. 72–98.

—— (2005a) *The Chinese State in Ming Society*, London and New York: Routledge Curzon.

—— (2005b) 'Introduction: a grave in Nanchang', in Timothy Brook, *The Chinese State in Ming Society*, London and New York: Routledge Curzon, pp. 1–14.

—— (2005c) 'Growing rice in North Zhili', Timothy Brook, *The Chinese State in Ming Society*, London and New York: Routledge Curzon, pp. 79–96.

Brumfiel, Elizabeth M. (2006) 'Cloth, gender, continuity, and change: fabricating unity in anthropology', AAA Presidential Address, *American Anthropologist* 108, 4: 862–877.

Campbell, Cameron D., Feng Wang and James Z. Lee (2002) 'Pretransitional fertility in China', *Population and Development Review* 28, 4: 735–750.

Carlitz, Katherine (1991) 'The social uses of female virtue in late Ming editions of *Lienü zhuan*', *Late Imperial China* 12, 2: 117–152.

—— (1994) 'Desire, danger, and the body: stories of women's virtue in late Ming China', in Christina K. Gilmartin, Gail Hershatter, Lisa Rofel and Tyrene Whites (eds), *Engendering China: Women, Culture, and the State*, Cambridge, MA: Harvard University Press, pp. 101–124.

Carsten, Janet and Stephen Hugh-Jones (eds) (1995) *About the House: Lévi-Strauss and Beyond*, Cambridge: Cambridge University Press.

Cartier, Michel (1984) 'Travail et idéologie dans la Chine antique', in Michel Cartier (ed.), *Le travail et ses représentations*, Paris: Éditions des archives contemporaines, pp. 275–304.

Chao Kang (1977) *The Development of Cotton Textile Production in China*, Cambridge, MA: Harvard University Press.

Chao Yuan-ling (2009) *Medicine and Society in Late Imperial China: A Study of Physicians in Suzhou, 1600–1850*, New York: Peter Lang Publishing.

Chase-Dunn, Christopher (2010) '*Adam Smith in Beijing*: A World-Systems Perspective', *Historical Materialism* 18: 39–51.

Chemla, Karine, Francesca Bray, Daiwie Fu, Yilong Huang and Georges Métailié (eds) (2001) *La storia della scienza in Cina*, in Sandro Petruccioli (ed.), *Enciclopedia internazionale della historia della scienza*, Rome, Istituto Treccani, Volume II: 1–608.

Chen Hengli and Wang Dacan (1983) *Bu nongshu jiaoyi* (Comparative modern edition of [*Master Shen's agricultural treatise* and] the *Supplemented agricultural treatise*), Beijing: Nongye Press.

Chen Zugui (ed.) (1957) *Mian* (Cotton), Shanghai: Zhonghua Press (Chinese Agricultural Heritage Series no. 5).

Cheung Sui-wai (2008) *The Price of Rice: Market Integration in Eighteenth-Century China*, Bellingham, WA: Center for East Asian Studies, Western Washington University.

—— (2014, in press) 'A desire to eat well: rice and the market in eighteenth-century China', in Francesca Bray, Peter Coclanis, Edda Fields-Black and Dagmar Schäfer (eds), *Rice: Global Networks and New Histories*, New York: Cambridge University Press.

Chia, Lucille (2002) *Printing for Profit: The Commercial Publishers of Jianyang, Fujian (11th–17th Centuries)*, Cambridge, MA: Harvard University Asia Center.

Cho, Philip S. (2008) 'Popular religion and artisanal practice in late imperial China: sericulture songs and the urbanization of silkwork temples in 18th century Jiangnan', paper presented at the panel on 'Religion in harmony with technology', *Society for the History of Technology* annual meeting, Lisbon, 11–14 October 2008.

Clément, Sophie, Pierre Clément and Shin Yong-hak (1987) *Architecture du paysage en Extrême-Orient*, Paris: Ecole nationale supérieure des Beaux-Arts.

Clunas, Craig (1991) *Superfluous Things: Material Culture and Social Status in Early Modern China*, Oxford: Polity Press.

—— (1996) 'Furnishing the self in early modern China', in Nancy Berliner (ed.), *Beyond the Screen: Chinese Furniture of the 16th and 17th Centuries*, Boston, MA: Museum of Fine Arts, pp. 21–35.

—— (1997) *Pictures and Visuality in Early Modern China*, Princeton, NJ: Princeton University Press.

—— (1998) 'Luxury knowledge: the *Xiushilu* [Records of Lacquering] of 1625', *Techniques et culture* 27 (Summer 1998): 37–40.

—— (1999) 'Modernity global and local: consumption and the rise of the West', *American Historical Review* 104, 5: 1497–1511.

Cohen, Myron (1976) *House United, House Divided: The Chinese Family in Taiwan*, Palo Alto, CA: Stanford University Press.

Corbin, Alain (1982) *Le miasme et la jonquille: l'odorat et l'imaginaire social, XVIIIe–XXe siècle*, Paris: Aubier-Montaigne.

Coupaye, Ludovic (2009) 'Ways of enchanting: *chaînes opératoires* and yam cultivation in Nyamikum Village, Maprik, Papua New Guinea', *Journal of Material Culture* 14, 4: 433–458.

Cowan, Ruth Schwartz (1983) *More Work for Mother: The Ironies of Household Technology from the Open Hearth to the Microwave*, New York: Basic Books.

Crowell, William G. (1990) 'Northern emigrés and the problems of census registration under the Eastern Jin and Southern Dynasties', in Albert E. Dien (ed.), *State and Society in Early Medieval China*, Palo Alto, CA: Stanford University Press, pp. 171–210.

Cullen, Christopher (1993) 'Patients and healers in late imperial China: evidence from the *Jinpingmei*', *History of Science* xxxi: 91–150.

—— (2001) 'Yi'an (case statements): the origins of a genre of Chinese medical literature', in Elisabeth Hsu (ed.), *Innovation in Chinese Medicine*, Cambridge: Cambridge University Press, pp. 297–323.

Daniels, Christian (1996) *Science and Civilisation in China*, Volume VI Part 3, *Biology and Biological Technology, Agro-industries: Sugarcane Technology*, Cambridge: Cambridge University Press.

Dardess, John W. (1989) 'A Ming landscape: settlement, land use, labor, and estheticism in T'ai-ho County, Kiangsi', *Harvard Journal of Asiatic Studes* 49 (December): 295–364.

Davis, Mike (1992) *City of Quartz*, New York: Vintage.

de Bary, William Theodore (ed.) (1960) *Sources of Chinese Tradition*, New York: Columbia Press.

de Certeau, Michel (1984) (tr. Steven Randall) *The Practice of Everyday Life*, Berkeley, CA: University of California Press.

de Laet, Marianne and Annemarie Mol (2000) 'The Zimbabwe bush pump: mechanics of a fluid technology', *Social Studies of Science* 30, 2: 225–263.

de Vries, Jan (1994) 'The industrial revolution and the industrious revolution', *Journal of Economic History* 54, 2: 249–270.
Dien, Albert E. (ed.) (1990) *State and Society in Early Medieval China*, Palo Alto, CA: Stanford University Press.
Dobres, Marcia-Ann (2010) 'Archaeologies of technology', *Cambridge Journal of Economics* 34: 103–114.
Duchesne, Ricardo (2005) 'Peer Vries, the Great Divergence and the California School: who's in and who's out?' *World History Connected*, May 2005, available at: http://worldhistoryconnected.press.illinois.edu/2.2/duchesne.html (accessed 14 February 2011).
Duden, Barbara (1991) *The Woman Beneath the Skin: A Doctor's Patients in Eighteenth-Century Germany*, Cambridge, MA: Harvard University Press.
Durkheim, Émile (1915) (tr. J.W. Swain) *The Elementary Forms of the Religious Life*, London: George Allen & Unwin.
Dutton, Michael, Sanjay Seth and Leela Gandhi (2002) 'Plumbing the depths: toilets, transparency and modernity', *Postcolonial Studies: Culture, Politics, Economy* 2 (July 2002): 137–142.
Ebrey, Patricia B. (1974) 'Estate and family management in the Later Han as seen in the *Monthly Instructions for the Four Classes of People*', *Journal of the Economic and Social History of the Orient* 17, 2: 173–205.
—— (1984) *Family and Property in Sung China: Yuan Ts'ai's Precepts for Social Life*, Princeton, NJ: Princeton University Press.
—— (1991a) *Chu Hsi's Family Rituals: a Twelfth-Century Chinese Manual for the Performance of Cappings, Weddings, Funeral, and Ancestral Rites*, Princeton, NJ: Princeton University Press.
—— (1991b) *Confucianism and Family Rituals in Imperial China*, Princeton, NJ: Princeton University Press.
—— (1991c) 'Shifts in marriage finance from the sixth to the thirteenth century', in Rubie S. Watson and Patricia B. Ebrey (eds), *Marriage and Inequality in Chinese Society*, Berkeley, CA: University of California Press, pp. 97–132.
—— (1993) *The Inner Quarters: Marriage and the Lives of Chinese Women in the Sung Period*, Berkeley, CA: University of California Press.
—— (2003) *Women and the Family in Chinese History*, London and New York: Routledge.
Edgerton, David (2007a) *The Shock of the Old: Technology and Global History Since 1900*, London: Profile; New York: Oxford University Press.
—— (2007b) 'Creole technologies and global histories: rethinking how things travel in time and space', *HoST* 1: 75–112.
Ellen, Roy (1986) 'Microcosm, macrocosm and the Nuaulu house: concerning the reductionist fallacy as applied to metaphorical levels', *Bijdragen tot de Taal-, Land- en Volkenkunde* 142, 1: 1–30.
Elias, Norbert (1982 [1939]) (tr. Edmund Jephcott) *The Civilizing Process* [*Über den Prozess der Zivilisation: Soziogenetische und Psychogenetische Untersuchungen*], Oxford: Basil Blackwell.
—— (1985 [1933]) (tr. Pierre Kamnitzer and Jeanne Etoré) *La société de cour*, Paris: Calmann-Lévy.
Elman, Benjamin (2005) *On Their Own Terms: Science in China, 1550–1900*, Cambridge, MA: Harvard University Press.
Elvin, Mark (1973) *The Pattern of the Chinese Past: A Social and Economic Interpretation*, Palo Alto, CA: Stanford University Press.
—— (2004) *The Retreat of the Elephants: An Environmental History of China*, New Haven, CT: Yale University Press.
—— (2008) 'The historian as haruspex', *New Left Review* 52: 83–109.

Elvin, Mark and Ts'ui-jung Liu (eds) (1998) *Sediments of Time: Environment and Society in Chinese History*, Cambridge: Cambridge University Press.

Etkin, Nina L. (1988) 'Cultural constructions of efficacy', in S. van der Geest and S. R. Whyte (eds), *The Context of Medicine in Developing Countries: Studies in Pharmaceutical Anthropology*, Dordrecht: Kluwer Academic Publishers, pp. 299–326.

Evarts, Curtis (1996) 'The furniture maker and the woodworking traditions of China', in Nancy Berliner (ed.), *Beyond the Screen: Chinese Furniture of the 16th and 17th Centuries*, Boston, MA: Museum of Fine Arts, pp. 53–76

Eyferth, Jacob (2009) *Eating Rice from Bamboo Shoots: The Social History of a Community of Handicraft Papermarkers in Rural Sichuan, 1920–2000*, Cambridge, MA: Harvard East Asia Monographs.

Faulkner, Philip, Clive Lawson and Jochen Runde (2010) 'Theorising technology', *Cambridge Journal of Economics* 34: 1–16.

Feenberg, Andrew (1999) *Questioning Technology*, New York: Routledge.

—— (2010) 'Marxism and the critique of social rationality: from surplus value to the politics of technology', *Cambridge Journal of Economics* 34: 37–49.

Ferguson, Eugene S. (1977) 'The mind's eye: nonverbal thought in technology', *Science* 197 (26 August): 827–836.

Feuchtwang, Stefan (1974) *An Anthropological Analysis of Chinese Geomancy*, Vientiane: Vithagna.

Finlay, Robert (2010) *The Pilgrim Art: Cultures of Porcelain in World History*, Berkeley, CA: University of California Press.

Flad, Rowan, Li Suicheng, Wu Xianhong and Zhao Zhijun (2010) 'Early wheat in China: results from new studies at Donghuishan in the Hexi Corridor', *The Holocene* 20: 955–965.

Flath, James A. (2003) *The Nianhua Gallery*, available at: http://history.uwo.ca/nianhua (accessed 14 July 2011).

Flitsch, Mareile (2008) 'Knowledge, embodiment, skill and risk: anthropological perspectives on women's everyday technologies in rural North China', in Francesca Bray (ed.), *East Asian Science, Technology and Society: An International Journal*, special issue, *Constructing Intimacy: Technology, Family and Gender in East Asia*, 2, 2, pp. 265–288.

Fong, Grace S. (2004) 'Female hands: embroidery as a knowledge field in women's everyday life in late imperial and early Republican China', *Late Imperial China* 25, 1: 1–58.

—— (2008) *Herself an Author: Gender, Agency and Writing in Late Imperial China*, Honolulu, HI: University of Hawaii Press.

Foucault, Michel (1973) *The Birth of the Clinic*, New York: Vintage.

Francks, Penelope (2014, in press) 'Rice and the path of economic development in Japan', in Francesca Bray, Peter Coclanis, Edda Fields-Black and Dagmar Schäfer (eds), *Rice: Global Networks and New Histories*, New York: Cambridge University Press.

Frank, André Gunder (1998) *ReOrient: Global Economy in the Asian Age*, Berkeley, CA: University of California Press.

Franke, Otto (1913) *Keng Tschi T'u: Ackerbau und Seidengewinnung in China*, Hamburg: L. Friederichsen & Co.

Freedman, Maurice (1969) 'Geomancy', *Proceedings of the Royal Anthropological Institute*: 5–15.

Freer Gallery of Art (2010a) *Song and Yuan Painting and Calligraphy, Poem Illustration and Genre Painting, 13, Cheng Qi (attr.), Tilling Rice after Lou Shou, F1954.21*, available at: www.asia.si.edu/SongYuan/F1954.21/F1954.21.asp (accessed 15 August 2012).

—— (2010b) *Song and Yuan Painting and Calligraphy, Poem Illustration and Genre Painting, 14, Cheng Qi (attr.), Silk Weaving after Lou Shou, F1954.20*, available at: www.asia.si.edu/SongYuan/F1954.20/F1954.20.asp (accessed 15 August 2012).

Furth, Charlotte (1987) 'Pregnancy, childbirth and infancy in Ch'ing dynasty China', *Journal of Asian studies* 46, 1: 7–35.
—— (1994) 'Rethinking van Gulik: sexuality and reproduction in traditional Chinese medicine', in Christina K. Gilmartin, Gail Hershatter, Lisa Rofel and Tyrene White (eds), *Engendering China: Women, Culture, and the State*, Cambridge, MA: Harvard University Press, pp. 125–146.
—— (1999) *A Flourishing Yin: Gender in China's Medical History, 960–1665*, Berkeley, CA: University of California Press.
—— (2007a) 'Introduction: thinking with cases', in Charlotte Furth, Judith T. Zeitlin, and Ping-chen Hsiung (eds), *Thinking with Cases: Specialist Knowledge in Chinese Cultural History*, Honolulu, HI: University of Hawaii Press, pp. 1–27.
—— (2007b) 'Producing medical knowledge through cases: history, evidence, and action', in Charlotte Furth, Judith T. Zeitlin, and Ping-chen Hsiung (eds), *Thinking with Cases: Specialist Knowledge in Chinese Cultural History*, Honolulu, HI: University of Hawaii Press, pp. 125–151.
Furth, Charlotte, Judith T. Zeitlin, and Ping-chen Hsiung (eds) (2007) *Thinking with Cases: Specialist Knowledge in Chinese Cultural History*, Honolulu, HI: University of Hawaii Press.
Gao Hanyu (1986) *Zhongguo lidai zhi ran xiu tulu*, Hong Kong: Commercial Press. English edn., (1992) (tr. Rosemary Scott and Susan Whitfield) *Chinese Textile Designs*, London: Viking/Penguin.
Gao Yinxian and Yi Nianhua (ed. Gong Zhebing) (1991) *Nüshu – shijie weiyi di nüxing wenzi* (Women's script: the world's only female script), Taipei: Funü xinzhiji jinhui Press.
Gardella, Robert (1994) *Harvesting Mountains: Fujian and the China Tea Trade, 1757–1937*, Berkeley, CA: University of California Press.
Gates, Hill (1989) 'The commoditization of Chinese women', *Signs* 14, 4: 799–832.
—— (1996) *China's Motor: A Thousand Years of Petty Capitalism*, Ithaca, NY: Cornell University Press.
Geertz, Clifford (1963) *Agricultural Involution: The Processes of Ecological Change in Indonesia*, Berkeley, CA: University of California Press.
Giddens, Anthony (1984) *The Constitution of Society: Outline of the Theory of Structuration*, Berkeley, CA: University of California Press.
Giedion, Siegfried (1948) *Mechanization Takes Command: A Contribution to Anonymous History*, New York: Oxford University Press.
Gilmartin, Christina K., Gail Hershatter, Lisa Rofel and Tyrene Whites (eds) (1994) *Engendering China: Women, Culture, and the State*, Cambridge, MA: Harvard University Press.
Golas, Peter J. (1980) 'Rural China in the Song', *Journal of Asian Studies* 39, 2: 291–325.
—— (1999) 'The emergence of technical drawing in China; the *Xin yi xiang fa yao* and its antecedents', *History of Technology* 21: 29–63.
—— (2001) 'Technological illustration in China; a post-Needham perspective', in Alain Arrault and Catherine Jami (eds), *Science and Technology in East Asia: The Legacy of Joseph Needham*, Turnhout, Belgium: Brepols, pp. 43–58.
—— (2007) '"Like obtaining a great treasure": the illustrations in Song Yingxing's *The Exploitation of the Works of Nature*', in Francesca Bray, Vera Dorofeeva-Lichtmann and Georges Métailié (eds), *Graphics and Text in the Production of Technical Knowledge in China*, Leiden: Brill, pp. 569–614.
Goldstone, Jack A. (2000) 'The rise of the West – or not? A revision to socio-economic history', *Sociological Theory* 18, 2: 175–194.
—— (2002) 'Efflorescences and economic growth in world history: rethinking the "rise of the West" and the Industrial Revolution', *Journal of World History* 13, 2: 323–389.

Goody, Esther N. (1982) 'Introduction', in Esther N. Goody (ed.), *From Craft to Industry: The Ethnography of Proto-Industrial Cloth Production*, Cambridge: Cambridge University Press, pp. 1–37.

Grant, Joanna (2003) *A Chinese Physician: Wang Ji and the 'Stone Mountain Medical Case Histories'*, London: Routledge Curzon.

Grapard, Allan G. (1994) 'Geosophia, geognosis and geopiety: orders of significance in Japanese representations of space', in Roger Friedland and Deirdre Boden (eds), *NowHere: Space, Time and Modernity*, Berkeley, CA: University of California Press, pp. 372–401.

Grist, D.H. (1975) *Rice*, 5th edn., London: Longman.

Hafter, Daryl M. (2007) *Women at Work in Preindustrial France*, University Park, PA: Penn State University Press.

Hall, Catherine (1990) 'The sweet delights of home', in Michelle Perrot (ed.) (tr. Arthur Goldhammer), *The History of Private Life*, Volume IV, *From the Fires of Revolution to the Great War*, Cambridge, MA: Belknap Press of Harvard University, pp. 47–93.

Hamilton, Gary G. and Wei-An Chang (2003) 'The importance of commerce in the organization of China's late imperial economy', in Giovanni Arrighi, Takeshi Hamashita and Mark Selden (eds), *The Resurgence of East Asia: 500, 150 and 50 Year Perspectives*, London and New York: Routledge, (eds), pp. 173–213.

Hammers, Roslyn (2005) '"How can it be that a crow's tail can hold water?" The square-pallet pump in Lou Shu's "Pictures of Tilling and Weaving"', *Technology and Culture* 46, 1: 132–136.

—— (2011) *Pictures of Tilling and Weaving: Art, Labor and Technology in Song and Yuan China*, Hong Kong: Hong Kong University Press.

Hanan, Patrick (1988) *The Invention of Li Yu*, Cambridge, MA: Harvard University Press.

Handlin, Joanna F. (1975) 'Lü Kun's new audience: the influence of women's literacy on sixteenth-century thought', in Margery Wolf and Roxane Witke (eds), *Women in Chinese Society*, Palo Alto, CA: Stanford University Press, pp. 13–38.

Harvey, David (1990) *The Condition of Postmodernity*, Oxford: Blackwell.

Harwood, Jonathan (2005) *Technology's Dilemma*, London: Peter Lang.

Hayami Akira (1967) '*Keizai shakai no seiritsu to sono tokushitsu*' (The emergence of economic society and its characteristics), in Shakai Keizaishi Gakkai (ed.), *Atarashii Edo Jidaizo o Motomete* (In search of a new image of the Edo Period), Tokyo: Toyo Keizai Shinposha.

—— (1992) 'The Industrious Revolution', *Look Japan* 38, 436: 8–10.

Hayden, Dolores (1986) *Redesigning the American Dream: The Future of Housing, Work and Family Life*, New York: Norton.

He Xiaoxin (1990) *Fengshui tanyuan* (Exploring the sources of *fengshui*), Nanjing: Dongnan University Press.

Hegel, Robert E. (1998) *Reading Illustrated Fiction in Late Imperial China*, Palo Alto, CA: Stanford University Press.

Heller, Chaia (2007) 'Techne versus technoscience: divergent (and ambiguous) notions of food "quality" in the French debate over GM crops', *American Anthropologist* 109, 4: 603–615.

Herlihy, David (1985) *Medieval Households*, Cambridge, MA: Harvard University Press.

Ho Ping-ti (1956) 'Early ripening rice in Chinese history', *Economic History Review* 9, 2: 200–218.

Horowitz, Robert (ed.) (2001) *Boys and Their Toys? Masculinity, Class, and Technology in America*, New York and London: Routledge.

Hostetler, Laura (2001) *Qing Colonial Enterprise: Ethnography and Cartography in Early Modern China*, Chicago, IL: University of Chicago Press.

Hsiung Ping-chen (2005) *A Tender Voyage: Children and Childhood in Late Imperial China*, Palo Alto, CA: Stanford University Press.

—— (2007) 'Facts in the tale: case records and pediatric medicine in late imperial China', in Charlotte Furth, Judith T. Zeitlin, and Ping-chen Hsiung (eds), *Thinking with Cases: Specialist Knowledge in Chinese Cultural History*, Honolulu, HI: University of Hawaii Press, pp. 152–168.

Hsu Cho-yun (ed. Jack Dull) (1980) *Han Agriculture: The Formation of the Early Chinese Agrarian Economy, 206 B.C. – A.D. 220*, Seattle, WA: University of Washington Press.

Hu Daojing (ed.) (1963) *Zhongyi biyong* (Practices necessary for growing crops), Beijing: Agriculture Press.

—— (1980) 'Xu Guangqi yanjiu nongxue licheng di tansuo' (On the dating of Xu Guangqi's agronomic research), *Lishi yanjiu* 6: 117–134.

—— (1985) *Nongshu, nongshi lunji* (Collected essays on agricultural writing and agricultural history), Beijing: Nongye Press.

Huang I-fen (2012) 'Gender, technical innovation, and Gu Family embroidery in late-Ming Shanghai', *EASTM (East Asian Science, Technology and Medicine)* 36, in press.

Huang, Philip C. C. (1985) The Peasant Economy and Social Change in North China, Palo Alto, CA: Stanford University Press.

—— (1990) *The Peasant Family and Rural Development in the Yangzi Delta, 1350–1988*, Palo Alto, CA: Stanford University Press.

Huang, Ray (1974) *Taxation and Government Finance in 16th Century Ming China*, Cambridge: Cambridge University Press.

Huang Shengwu (ed.) (1983) *Zhongguo yixue baike quanshu. Zhongyi fukexue* (Chinese medical encyclopedia. Traditional Chinese gynecology), Shanghai: Xinhua Press.

Hughes, Thomas P. (1983) *Networks of Power: Electrification in Western Society 1880–1930*, Baltimore, MD: Johns Hopkins University Press.

Hutchins, Edwin (1995) *Cognition in the Wild*, Cambridge, MA: MIT Press.

Ingold, Tim (2000) *The Perception of the Environment: Essays in Livelihood, Dwelling and Skill*, London: Routledge.

—— (2007) 'Materials against materiality', *Archaeological Dialogues* 14, 1: 1–18.

Jami, Catherine, Peter Engelfriet and Gregory Blue (eds) (2001) *Statecraft and Intellectual Renewal in Late Ming China: the Cross-Cultural Synthesis of Xu Guangqi (1562–1633)*, Leiden: Brill.

Janousch, Andreas (2011) 'The censor's stele: religion, salt-production and labour in the temple of the God of the Salt Lake in Southern Shanxi Province', paper presented at the workshop on 'Ritual as technology in East Asia', organized by Grace Shen, Max Planck Institute for the History of Science, Berlin, 12–14 May 2011.

Jin Hansheng (1988) 'Songdai nüzi zhiye yu shengji' (Female occupations and livelihoods in the Song), in Bao Jialin (ed.), *Zhongguo funü shi lun ji* (Collected essays on the history of Chinese women), Beijing: Daoxiang Press, pp. 193–204.

Johnson, Elizabeth (1977) 'Patterned bands in the New Territories of Hong Kong', *Journal of the Royal Asiatic Society, Hong Kong Branch* 17: 81–91.

Jones, Eric L. (1981) *The European Miracle: Environments, Economies, and Geopolitics in the History of Europe and Asia*, Cambridge: Cambridge University Press.

Kendall, Laurel (1996) 'Korean shamans and the spirits of capitalism', *American Anthropologist* 98, 3: 512–527.

Kloppenburg, Jack Ralph JR. (1988) *First the Seed: The Political Economy of Plant Biotechnology, 1492–2000*, Cambridge: Cambridge University Press.

Knapp, Ronald G. (1998) *China's Living Houses: Folk Beliefs, Symbols, and Household Ornamentation*, Honolulu, HI: University of Hawaii Press.

Knapp, Ronald G. and Kai-Yin Lo (eds) (2005) *House Home Family: Living and Being Chinese*, Honolulu, HI: University of Hawaii Press.

Ko, Dorothy (1993) *Teachers of the Inner Chambers: Women and Culture in Seventeenth-Century China*, Palo Alto, CA: Stanford University Press.

—— (1994) *Teachers of the Inner Chambers: Women and Culture in Seventeenth-Century China*, Palo Alto, CA: Stanford University Press.

—— (2005) *Cinderella's Sisters: A Revisionist History of Foot-Binding*, Berkeley, CA: University of California Press.

Köhle, Natalie (2008) 'Why did the Kangxi Emperor go to Wutai Shan? Patronage, pilgrimage, and the place of Tibetan Buddhism at the early Qing court', *Late Imperial China* 29, 1: 73–119.

Kuhn, Dieter (1976) 'Die Darstellung des *Keng-chih-t'u*', *Zeitschrift der Deutschen Morgenländischen Gesellschaft* 126, 2: 336–367.

—— (1977) 'Marginalie zu einigen Illustrationen im Nung-shu des Wang Chen', in *Zur Kunstgeschichte Asiens: 50 Jahre Lehre und Forschung an der Universität Köln*, Wiesbaden: Franz Steiner, pp. 143–152.

—— (1987) *Die Song-Dynastie (960 bis 1279): eine neue Gesellschaft im Spiegel ihrer Kultur*, Wienheim: Acta Humaniorum VCH.

—— (1988) *Science and Civilisation in China*, Volume V Part 9, *Textile Technology Part I: Spinning*, Cambridge: Cambridge University Press.

—— (2006) 'Reflections on the current state and significance of the history of East Asian technology', *East Asian Science, Technology and Medicine* 25: 9–26.

—— (2009) *The Age of Confucian Rule: The Song Transformation of China*, Cambridge MA: Harvard University Press.

Lackner, Michael (2000) 'Was Millionen Wörter nicht sagen können: Diagramme zur Visualisierung klassischer Texte im China des 13. bis 14. Jahrhunderts', *Semiotik* 22, 2: 209–237.

Landes, David S. (1969) *The Unbound Prometheus: Technological Change and Industrial Development in Western Europe from 1750 to the Present*, Cambridge: Cambridge University Press.

Langlois, John D. (2001) 'Review of *A Cultural History of Civil Examinations in Late Imperial China by Benjamin A. Elman*', *Harvard Journal of Asiatic Studies* 61, 1: 216–230.

Lansing, J. Stephen (2006) *Perfect Order: Recognizing Complexity in Bali*, Princeton, NJ: Princeton University Press.

Latour, Bruno (1986) 'Visualization and cognition: thinking with eyes and hands', *Knowledge and Society: Studies in the Sociology of Culture Past and Present*, 6: 1–40.

—— (1987) *Science in Action: How to Follow Scientists and Engineers through Society*, Cambridge MA: Harvard University Press.

—— (1991) 'Technology is society made durable', in John Law (ed.), *A Sociology of Monsters: Essays on Power, Technology and Domination*, London: Routledge, pp. 103–131.

—— (1992) 'Where are the missing masses? The sociology of a few mundane artefacts', in Wiebe Bijker and John Law (eds), *Shaping Technology/Building Society: Studies in Sociotechnical Change*, Cambridge, MA: MIT Press, pp. 225–258.

Lattimore, Owen (1962) *Studies in Frontier History*, London and New York: Oxford University Press.

Lavely, William, and R. Bin Wong (1998) 'Revising the Malthusian narrative: the comparative study of population dynamics in late imperial China', *Journal of Asian Studies* 57, 3: 714–748.

Lavely, William, James Lee and Feng Wang (1990) 'Chinese demography: the state of the field', *Journal of Asian Studies* 49, 4: 807–834.

Law, John (1986) 'On the methods of long distance control: vessels, navigation and the Portuguese route to India', in John Law (ed.), *Power, Action and Belief: A New Sociology of Knowledge?*, London: Routledge and Kegan Paul, pp. 234–263.

Ledderose, Lothar (2000) *Ten Thousand Things: Module and Mass Production in Chinese Art*, Princeton, NJ: Princeton University Press.

Lee, James, Cameron Campbell, and Feng Wang (2002) 'Positive check or Chinese checks?', *Journal of Asian Studies* 61, 2: 591–607.

Lee, James and Feng Wang (1999) *One Quarter of Humanity: Malthusian Mythology and Chinese Realities, 1700–2000*, Cambridge, MA: Harvard University Press.

Lee Seung-Joon (2014, in press) 'Rice and maritime modernity: the modern Chinese state and the South China Sea rice trade', in Francesca Bray, Peter Coclanis, Edda Fields-Black and Dagmar Schäfer (eds), *Rice: Global Networks and New Histories*, New York: Cambridge University Press.

Lefebvre, Henri (1991) (tr. Donald Nicholson-Smith) *The Production of Space*, Oxford: Blackwell, 1991 (1st edn., *La production de l'espace*, Paris: Editions Anthropos, 1974).

Lefèvre, Wolfgang (ed.) (2004) *Picturing Machines, 1400–1700*, Cambridge, MA: MIT Press.

Lemonnier, Pierre (1992) *Elements for an Anthropology of Technology*, Ann Arbor, MI: Museum of Anthropology, University of Michigan.

—— (ed.) (1993) *Technological Choices: Transformation in Material Cultures since the Neolithic*, London: Routledge.

Lerman, Nina E. (2010) 'Categories of difference, categories of power: bringing gender and race to the history of technology', *Technology and Culture* 51, 4: 892–918.

Lerman, Nina E., Ruth Oldenziel and Arwen P. Mohun (eds) (2003) *Gender & Technology: A Reader*, Baltimore, MD: Johns Hopkins University Press.

Li Bozhong (1990) *Tangdai Jiangnan nongye di fazhan* (The development of agriculture in Tang-dynasty Jiangnan), Beijing: Nongye Press.

—— (1994) 'Kongzhi zengchang, yi bao fuyu – Qingdai qianzhongqi Jiangnan de renkou xingwei' (Controlling growth to protect wealth – population behavior in Jiangnan during the early and mid Qing period), *Xin shixue* 5, 3: 25–70.

—— (1998a) 'Changes in climate, land, and human efforts: the production of wet-field rice in Jiangnan during the Ming and Qing dynasties', in Mark Elvin and Ts'ui-jung Liu (eds), *Sediments of Time: Environment and Society in Chinese History*, Cambridge: Cambridge University Press, pp. 447–486.

—— (1998b) *Agricultural Development in Jiangnan, 1620–1850*, New York: St Martin's Press.

—— (2000) 'Duotai, biyun, yu jueyu: Song Yuan Ming Qing Jiangzhe diqu de jieyu fangfa jiqi yunyong chuanbo' (Abortion, contraception and sterilization: birth control methods and their dissemination in Song-Yuan-Ming-Qing Jiangsu and Zhejiang), in James Lee, Songyi Guo and Yizhuang Ding (eds), *Hunying, jiating yu renkou xingwei: dongxi bijiao* (Marriage, family formation and population behavior: East-West comparisons), Beijing: Beijing University Press, pp. 71–99.

Li Changnian (1982) *Nongsang jing jiaozhu* (Collated and annotated edition of the Classic of Farming and Sericulture), Beijing: Nongye Press.

—— (1983) 'Xu Guangqi di nongzheng sixiang' (The agronomic thought of Xu Guangqi), *Zhongguo nongshi* (Chinese agricultural history) 3: 1–8.

Li Cho-ying (2010) 'Contending strategies, collaboration among local specialists and officials, and hydrological reform in the late fifteenth-century Lower Yangzi Delta', *East Asian Science, Technology and Society* 4, 2: 229–253.

—— (2012) 'Beneficiary pays: forging reciprocal connections between private profit and public good in hydraulic reform in the Lower Yangzi Delta, 1520s–1640s', *T'oung Pao* 98: 385–438.

Li Jingwei (ed.) (1988) *Zhongyi renwu cidian* (Dictionary of Chinese physicians), Shanghai: Shanghai Zishu Press.

Li, Lillian M. (2007) *Fighting Famine in North China: State, Market, and Environmental Decline, 1690s–1990s*, Palo Alto, CA: Stanford University Press.

Li Yuhang (2012) 'Embroidering Guanyin: constructions of the divine through thread and hair', *EASTM (East Asian Science, Technology and Medicine)* 36, in press.

Liang Jiamian (1963) '*Nongzheng quanshu* zhuanshu guocheng ji ruogan youguan wenti de tantao' (A discussion of the processes of compilation of the *Nongzheng quanshu* and some related problems), in Zhongguo kexueyuan Ziranshi shi (ed.), *Xu Guangqi jinian lunwen* (Collected essays in memory of Xu Guangqi), Shanghai: Shanghai Zhonghua shuju, pp. 78–109.

Liu, Jerry and Kent Deng (eds) (2009) *History of Technology* 29 (special issue, *Chinese Technological History: The Great Divergence*).

Liu, Lydia H. (1999) 'Robinson Crusoe's earthenware pot', *Critical Inquiry* 25, 4: 728–757.

Lohan, Maria and Wendy Faulkner (2004) 'Masculinities and technologies: some introductory remarks', *Men and Masculinities* 6, 4 (special issue: *Masculinities and Technologies*): 319–329.

Long, Pamela O. (2007) 'The transmission of craft knowledge in Western Europe: methodological issues as a framework for comparison', paper presented at the workshop 'From invention to innovation', Max Planck Institute for the History of Science, Berlin, July 2007.

Low, Setha (2003) *Behind the Gates: Life, Security and the Pursuit of Happiness in Fortress America*, New York: Routledge.

Lu Houyuan, Jianping Zhang, Kam-biu Liu, Naiqin Wu, Yumei Li, Kunshu Zhou, Maolin Ye, Tianyu Zhang, Haijiang Zhang, Xiaoyan Yang, Licheng Shen, Deke Xu, Quan Li and Dolores R. Piperno (eds) (2009) 'Earliest domestication of common millet (Panicum miliaceum) in East Asia extended to 10,000 years ago', *Proceedings of the National Academy of Sciences of the United States of America (PNAS)* 106, 18: 7367–7372.

Lu Weijing (2004) 'Beyond the paradigm: tea-picking women in imperial China', *Journal of Women's History* 15, 4: 19–46.

Lu Yuanting (ed.) (1991–1996) *Zhongguo chuantong minju yu wenhua* (China's traditional vernacular dwellings and culture), 4 vols, Beijing: Zhongguo Jianzhu Gongye Press.

Luk, Bernard H.K. (1977) 'Abortion in Chinese law', *American Journal of Comparative Law* 25: 372–392.

McDermott, Joseph P. (1990) 'The Chinese domestic bursar', *Ajia bunka kenkyu*, pp. 15–32.

McMahon, Daniel (2009) 'Qing reconstruction in the Southern Shaanxi Highlands: state perceptions and plans, 1799–1820', *Late Imperial China* 30, 1: 85–118.

Mahias, Marie-Claude (1993) 'Pottery techniques in India: technical variants and social choice', in Pierre Lemonnier, *Elements for an Anthropology of Technology*, Ann Arbor, MI: Museum of Anthropology, University of Michigan,, pp. 157–180.

—— (2002) *Le barattage du monde. Essais d'anthropologie des techniques en Inde*, Paris: Éditions de la Maison des sciences de l'homme.

Mair, Victor (1985) 'Language and ideology in the written popularizations of the *Sacred Edict*', in David Johnson, Andrew J. Nathan and Evelyn S. Rawski (eds), *Popular Culture in Late Imperial China*, Berkeley, CA: University of California Press, pp. 325–359.

Malinowski, Bronislaw (1948 [1925]) *Magic, Science and Religion and Other Essays*, Boston, MA: Beacon Press.

Mandel, Ernest (1975) *Late Capitalism*, London: Humanities Press.

Mann, Susan (1991) 'Grooming a daughter for marriage: brides and wives in the mid-Ch'ing period', in Rubie S. Watson and Patricia B. Ebrey (eds), *Marriage and Inequality in Chinese Society*, Berkeley, CA: University of California Press, pp. 204–230.

—— (1992a) 'Household handicrafts and state policy in Qing times', in Jane Kate Leonard and John R. Watt (eds), *To Achieve Security and Wealth: The Qing Imperial State and the Economy 1644–1911*, Ithaca, NY: Cornell University East Asia Program, pp. 75–95.

—— (1992b) 'Women's work in the Ningbo area, 1900–1936', in Thomas G. Rawski and Lillian M. Li (eds), *Chinese History in Economic Perspective*, Berkeley, CA: University of California Press, pp. 243–270.

—— (1994) 'The education of daughters in the mid-Ch'ing period', in Benjamin Elman and Alexander Woodside (eds), *Education and Society in Late Imperial China, 1600–1900*, Berkeley, CA: University of California Press, pp. 19–49.

—— (1997) *Precious Records: Women in China's Long Eighteenth Century*, Palo Alto, CA: Stanford University Press.

Marchant, S. (2003) *Recent Acquisitions 2003*, 120 Kensington Church Street, London.

Marks, Robert B. (1998a) *Tigers, Rice, Silk and Silt: Environment and Economy in Late Imperial South China*, Cambridge: Cambridge University Press.

—— (1998b) '"It never used to snow": climatic variability and harvest yields in late-imperial South China, 1650–1850', in Mark Elvin and Ts'ui-jung Liu (eds), *Sediments of Time: Environment and Society in Chinese History*, Cambridge: Cambridge University Press, pp. 411–446.

Martin, Emily (1988) 'Gender and ideological differences in representations of life and death', in James L. Watson and Evelyn S. Rawski (eds), *Death Ritual in Late Imperial and Modern China*, Berkeley, CA: University of California Press, pp. 203–226.

Marx, Leo (2010) 'Technology: the emergence of a hazardous concept', *Technology and Culture* 51: 561–577.

Mauss, Marcel (1950 [1936]) 'Les techniques du corps', in *Sociologie et anthropologie*, Paris, PUF: 365–386.

Mazumdar, Sucheta (1998) *Sugar and Society in China: Peasants, Technology, and the World Market*, Cambridge, MA: Harvard University Asia Center.

Mellström, Ulf (2002) 'Patriarchal machines and masculine embodiment', *Science, Technology and Human Values* 27: 460–478.

Menzies, Nicholas K. (1994) *Forest and Land Management in Imperial China*, London: Palgrave Macmillan.

—— (1996) *Science and Civilisation in China*, Volume VI Part 3, *Biology and Biological Technology, Agro-industries and Forestry*, Cambridge: Cambridge University Press.

Miao Qiyu (ed.) (1981a) *Chen Fu Nongshu xuandu* (Selected readings from Chen Fu's *Nongshu*), Beijing: Nongye Press.

—— (1981b) *Sishi zuanyao jiaoshi* (Annotated modern translation of the *Sishi zuanyao*), Beijing, Nongye Press.

Miao Qiyu and Guilong Miao (eds) (1982) *Qimin yaoshu jiaoshi* (Annotated modern translation of the *Qimin yaoshu*), Beijing: Nongye Press.

Miller, Daniel (2005) 'Materiality: an introduction', in Daniel Miller (ed.), *Materiality*, Durham, NC: Duke University Press, pp. 1–50.

Miller, Daniel and Don Slater (2000) *The Internet: An Ethnographic Approach*, Oxford and New York: Berg.

—— (1988) *Feminism and Anthropology*, Oxford, Polity Press and Basil Blackwell.

Mokyr, Joel (1990) *The Lever of Riches: Creativity and Economic Progress*, New York: Oxford University Press.

—— (2009) *The Enlightened Economy: An Economic History of Britain 1700–1850*, New Haven: Yale University Press.

Moll-Murata, Christine, Song Jianze, and Hans Ulrich Vogel (eds) (2005) *Chinese Handicraft Regulations of the Qing Dynasty: Theory and Application*, Munich: Iudicum.

Moon, Suzanne (2007) *Technology and Ethical Idealism: A History of Development in the Netherlands East Indies*, Amsterdam: Amsterdam University Press.

Moore, Henrietta L. (1986) *Space, Text and Gender: An Anthropological Study of the Marakwet of Kenya*, Cambridge: Cambridge University Press.

Mukerji, Chandra (2009) *Impossible Engineering: Technology and Territoriality on the Canal du Midi*, Princeton, NJ: Princeton University Press.

Mumford, Lewis (1934) *Technics and Civilization*, New York: Harcourt Brace.

Naji, Myriem and Laurence Douny (2009) 'Editorial', *Journal of Material Culture* 14, 4 (special issue on *"Making" and "doing" the material world*): 411–432.

Needham, Joseph (1969) *Clerks and Craftsman in China and the West: Lectures and Addresses on the History of Science and Technology*, Cambridge: Cambridge University Press.

—— (2000) 'Introduction', in Edgar Zilsel, Diederick Raven, Wolfgang Krohn and Robert S. Cohen (eds), *The Social Origins of Modern Science*, Dordrecht and Boston, MA: Kluwer Academic Publishers, pp. xi–xiv.

Needham, Joseph *et al.* (1954–) *Science and Civilisation in China*, Cambridge: Cambridge University Press.

Nichter, Mark (1992) 'Of ticks, kings, spirits, and the promise of vaccines', in Charles Leslie and Allen Young (eds) *Paths to Asian Medical Knowledge*, Berkeley, CA: University of California Press, pp. 224–255.

O'Brien, Patrick K. (2009) 'The Needham Question updated: a historiographical survey and elaboration', in Jerry Liu and Kent Deng (eds), *History of Technology* 29 (special issue, *Chinese Technological History: The Great Divergence*), pp. 7–28.

Osborne, Anne (1998) 'Highlands and lowlands: economic and ecological interactions in the lower Yangzi region under the Qing', in Mark Elvin and Ts'ui-jung Liu (eds), *Sediments of Time: Environment and Society in Chinese History*, Cambridge: Cambridge University Press, pp. 203–234.

Ou Ming (1988) *Chinese-English Dictionary of Traditional Medicine*, Hong Kong: Joint Publishing Press.

Pan Jixing (1989) *Tiangong kaiwu jiaozhu yu yanjiu* (Critical edition and study of the *Tiangong kaiwu*), Chengdu: Bashu shushe.

Pelliot, Paul (1913) *A propos du 'Keng Tche T'ou'*, Paris: Leroux, pour l'Académie des Inscriptions et Belles-Lettres.

Perdue, Peter C. (1987) *Exhausting the Earth: State and Peasant in Hunan, 1500–1850*, Cambridge, MA: Harvard University Press.

Perkins, Dwight H. (1968) *Agricultural Development in China, 1368–1968*, Chicago, IL: Aldine.

Perrot, Michelle (ed.) (1990) (tr. Arthur Goldhammer) *A History of Private Life, IV: From the Fires of the Revolution to the Great War*, Cambridge MA: Harvard, Belknap Press.

Pfaffenberger, Bryan P. (1992) 'Social anthropology of technology', *Annual Review of Anthropology* 21: 491–516.

—— (2001) 'Symbols do not create meanings – activities do: or, why symbolic anthropology needs the anthropology of technology', in Michael B. Schiffer (ed.), *Anthropological Perspectives on Technology*, Albuquerque, NM: University of New Mexico Press, pp. 77–86.

Pinch, Trevor J. and Wiebe E. Bijker (1987) 'The social construction of facts and artefacts', in Wiebe E. Bijker, Thomas P. Hughes and Trevor J. Pinch (eds), *The Social Construction of Technological Systems*, Cambridge, MA: MIT Press, pp. 17–51.

Pinney, Chris (2002) 'Visual culture', in Victor Buchli (ed.), *The Material Culture Reader*, Oxford: Berg, pp. 81–104.

Pomeranz, Kenneth (2000) *The Great Divergence: China, Europe, and the Origins of the Modern World Economy*, Princeton, NJ: Princeton University Press.

—— (2002) 'Political economy and ecology on the eve of industrialization: Europe, China, and the global conjuncture', *The American Historical Review* 107, 2: 425–446.

—— (2003) 'Women's work, family, and economic development in Europe and East Asia: long-term trajectories and contemporary comparisons', in Giovanni Arrighi, Takeshi Hamashita and Mark Selden (eds), *The Resurgence of East Asia: 500, 150 and 50 Year Perspectives*, London and New York: Routledge, pp. 124–172.

Pu Songling (2006) (tr. John Minford) *Strange Tales from a Chinese Studio*, Harmondsworth: Penguin Classics.

Puffert, Douglas (2008) 'Path dependence', *EH.Net Encyclopedia*, edited by Robert Whaples, 10 February 2008, available at: http://eh.net/encyclopedia/article/puffert.path.dependence (accessed 10 July 2011).

Pursell, Carroll (1994) *White Heat: People and Technology*, London: BBC Books.

Raphals, Lisa (1998) *Sharing the Light: Representations of Women and Virtue in Early China*, Albany, NY: SUNY Press.

Rawski, Evelyn S. (1972) *Agricultural Change and the Peasant Economy of South China*, Cambridge, MA: Harvard University Press.

Roberts, Lissa (2007) 'Introduction', in Lissa Roberts, Simon Schaffer and Peter Dear (eds), *The Mindful Hand: Inquiry and Invention from the Late Renaissance to Industrialisation*, Amsterdam: Royal Netherlands Academy of Arts and Sciences/KNAW, pp. 1–11.

Roberts, Lissa, Simon Schaffer and Peter Dear (2007) 'How to read this book', in Lissa Roberts, Simon Schaffer and Peter Dear (eds), *The Mindful Hand: Inquiry and Invention from the Late Renaissance to Industrialisation*, pp. ix–x.

Roberts, Robert (1971) *The Classic Slum: Salford in the First Quarter of the Century*, Manchester: University of Manchester Press.

Robertson, A.F. (1991) *Beyond the Family: The Social Organization of Human Reproduction*, Berkeley, CA: University of California Press.

Rowe, William T. (2001) *Saving the World: Chen Hongmou and Elite Consciousness in Eighteenth-Century China*, Palo Alto, CA: Stanford University Press.

Ruitenbeek, Klaas (1993) *Carpentry and Building in Late Imperial China: A Study of the Fifteenth-Century Carpenter's Manual Lu Ban Jing*, Leiden: E.J. Brill.

Rybczynski, Witold (1986) *Home: A Short History of an Idea*, Harmondsworth: Penguin.

Sabban, Françoise (1993) '"Suivre le temps du ciel": économie ménagère et gestion du temps dans la Chine du VIe siècle', in Maurice Aymard, Claude Grignon and Françoise Sabban (eds), *Le temps de manger: alimentation, emploi du temps et rythmes sociaux*, Paris: Editions Maison des Sciences de l'Homme/Institut National de Recherche Agronomique, pp. 81–108.

—— (1994) 'L'industrie sucrière, le moulin à sucre et les relations sino-portugaises aux XIVe-XVIIIe siècles', *Annales Histoires, Sciences Sociales* 49, 4: 817–862.

Sacks, Karen (1974) 'Engels revisited: women, the organization of production, and private property', in Michelle Z. Rosaldo and Louise Lamphere (eds), *Woman, Culture and Society*, Palo Alto, CA: Stanford University Press, pp. 207–222.

Sand, Jordan (2004) *House and Home in Modern Japan: Architecture, Domestic Space, and Bourgeois Culture, 1880–1930*, Cambridge, MA: Harvard University Press.

Schäfer, Dagmar (Xie Feng) (ed.) (2010) *Gongting yu difang: shiqi zhi shiba shiji de jishu jiaoliu (The court and the localities: technological knowledge circulation in the 17th and 18th century)*, Beijing: Zijincheng Press.

—— (2011a) *The Crafting of the 10,000 Things: Knowledge and Technology in Seventeenth-Century China*, Chicago, IL: University of Chicago Press.

—— (2011b) 'Inscribing the artefact and inspiring trust: the changing role of markings in the Ming era', *East Asian Science, Technology & Society* 5, 2: 239–265.

—— (ed.) (2012) *Cultures of Knowledge: Technology in Chinese History*. Leiden: Brill.

Schlosser, Eric (2001) *Fast Food Nation*, New York: Houghton Mifflin.

Schneewind, Sarah (2006) *A Tale of Two Melons: Emperor and Subject in Ming China*, Indianapolis, IN: Hackett Publishing Co.

Scott, James C. (1998) *Seeing Like a State: How Certain Schemes to Improve the Human Condition Have Failed*, New Haven, CT: Yale University Press.

—— (2009) *The Art of Not Being Governed: An Anarchist History of Upland Southeast Asia*, New Haven, CT: Yale University Press.

Servanti, Silvano and Françoise Sabban (2002) (tr. Antony Shugaar) *Pasta: The Story of a Universal Food*, New York: Columbia University Press.

Shi Shenghan (1959) *On the Fan Sheng-chih Shu: An Agricultural Book Written by Fan Sheng-chih in -1st Century China*, Beijing: Science Press.

—— (1965) *Simin yueling jiaozhu* (Annotated commentary on the *Simin yueling*), Beijing: Zhonghua.

—— (1979) *Nongzheng quanshu* (The complete treatise on agricultural administration), Shanghai: Guji Press, 3 vols.

Shi Shenghan and Kang Chengyi (eds) (1982[1959]) *Bianmin tuzuan* (Collection of pictures for the convenience of the people), attr. Guang Fan. Repr., Beijing: Agriculture Press.

Shiba Yoshinobu (1998) 'Environment versus water control: the case of the Southern Hangzhou Bay area from the mid-Tang through the Qing', in Mark Elvin and Ts'ui-jung Liu (eds), *Sediments of Time: Environment and Society in Chinese History*, Cambridge: Cambridge University Press, pp. 135–164.

Sigaut, François (1985) 'More (and enough) on technology', *History and Technology* 2: 115–132.

Silber, Cathy (1994) 'From daughter to daughter-in-law in the women's script of Southern Hunan', in Christina K. Gilmartin, Gail Hershatter, Lisa Rofel and Tyrene Whites (eds), *Engendering China: Women, Culture, and the State*, Cambridge, MA: Harvard University Press, pp. 47–68.

Sivin, Nathan (1982) 'Why the Scientific Revolution did not take place in China – or didn't it?', *Chinese Science* 5: 45–66.

Smil, Vaclav (1993) *China's Environmental Crisis: An Inquiry into the Limits of National Development*, Armonk, NY: M.E. Sharpe.

Smith, Paul J. (1991) *Taxing Heaven's Storehouse: Horses, Bureaucrats and the Destruction of the Sichuan Tea Industry*, Cambridge MA: Harvard-Yenching Monograph Series.

Smith, Richard J. and D.W.Y. Kwok (eds) (1993) *Cosmology, Ontology and Human Efficacy: Essays in Chinese Thought*, Honolulu, HI: University of Hawaii Press.

So, Alvin Y. (1986) *The South China Silk District: Local Historical Transformation and World-System Theory*, Albany, NY: SUNY Press.

Song Yingxing (1936) *Tiangong kaiwu* (The exploitation of the works of nature), Shanghai: Commercial Press, *Guoxue jiben congshu jianbian* series.Stanley, Autumn (1995) *Mothers and Daughters of Invention: Notes for a Revised History of Technology*, 2nd edn., New Brunswick, NJ: Rutgers University Press.

Staudenmaier, John M. (1985) *Technology's Storytellers: Reweaving the Human Fabric*, Cambridge, MA: MIT Press.

Stockard, Janice E. (1989) *Daughters of the Canton Delta: Marriage Patterns and Economic Strategies in South China, 1860–1930*, Palo Alto, CA: Stanford University Press.

Strathern, Marilyn (1996) 'Cutting the network', *Journal of the Royal Anthropological Institute* 2: 517–535.

Sun E-Tu Zen and Shiou-Chuan Sun (trs) (1966) *T'ien-kung k'ai-wu: Chinese technology in the seventeenth century*, University Park, PA: Penn State University Press.

Tang Changru (1990) 'Clients and bound retainers in the Six Dynasties period', in Albert E. Dien (ed.), *State and Society in Early Medieval China*, Palo Alto, CA: Stanford University Press, pp. 110–138.

T'ien Ju-k'ang (1988) *Male Anxiety and Female Chastity: A Comparative Study of Ethical Values in Ming-Ch'ing Times*, Leiden: Brill.

Tong Shuye (1981) *Zhongguo shougongye shangye fazhan shi* (A history of the development of handicrafts and commerce in China), Jianan: Jilu Press.

Topley, Marjorie (1978) 'Marriage resistance in rural Kwangtung', in Arthur Wolf (ed.), *Studies in Chinese Society*, Palo Alto, CA: Stanford University Press, pp. 247–268.

Traweek, Sharon (1996) '*Kokusai*, *gaiatsu* and *bachigai*: Japanese physicists' strategies for moving into the international political economy of science', in Laura Nader (ed.), *Naked Science: Anthropological Inquiry into Boundaries, Power and Knowledge*, London: Routledge, pp. 174–197.

Tsur Nyok-Ching (1983 [1907]) (tr. Peter Schran) 'Forms of business in the city of Ningpo in China', *Chinese Sociology and Anthropology* 15, 4.

Tunstall, Alexandra (2012) 'Beyond categorization: Zhu Kerou's tapestry painting, *Butterfly and Camellia*', in *EASTM (East Asian Science, Technology and Medicine)* 36, in press.

Twitchett, D.C. (1963) *Financial Administration under the T'ang Dynasty*, Cambridge: Cambridge University Press.

Übelhör, Monika (1969) *Hsu Kuang-ch'i (1562–1633) und seine Einstellung zum Christentum*, published Ph.D. dissertation, University of Hamburg.

van der Leeuw, Sander (1993) 'Giving the potter a choice: conceptual aspects of pottery techniques', in Pierre Lemonnier, *Elements for an Anthropology of Technology*, Ann Arbor, MI: Museum of Anthropology, University of Michigan, pp. 238–288.

Veblen, Thorstein (1934 [1899]) *The Theory of the Leisure Class: An Economic Study of Institutions*, New York: Random House.

Vermeer, Eduard B. (1998) 'Population and ecology along the frontier in Qing China', in Mark Elvin and Ts'ui-jung Liu (eds), *Sediments of Time: Environment and Society in Chinese History*, Cambridge: Cambridge University Press, pp. 235–279.

Viraphol, Sarasin (1997) *Tribute and Profit: Sino-Siamese Trade, 1652–1853*, Cambridge, MA: Harvard University Press.

von Falkenhausen, Lothar (1993) *Suspended Music: Chime-Bells in the Culture of Bronze Age China*, Berkeley, CA: University of California Press.

Wagner, Donald B. (1993) *Iron and Steel in Ancient China*, Leiden: Brill.

—— (1997) *The Traditional Chinese Iron Industry and its Modern Fate*, Richmond, Surrey: Curzon Press.

—— (2007) 'Song Yingxing's illustrations of iron production', in Francesca Bray, Vera Dorofeeva-Lichtmann and Georges Métailié (eds), *Graphics and Text in the Production of Technical Knowledge in China*, Leiden: Brill, pp. 615–632.

—— (2008) *Science and Civilisation in China*, Volume 5 Part 11, *Ferrous Metallurgy*, Cambridge: Cambridge University Press.

Waley-Cohen, Joanna (1999) *The Sextants of Beijing: Global Currents in Chinese History*, New York: W.W. Norton & Company.

Wang Chaosheng (ed.) (1995) *Zhongguo gudai gengzhi tu* (Farming and weaving pictures in ancient China), Beijing: China Agriculture Press.

Wang Guangyao (2010) 'Qianlong gongting li di Suzhou yugong' (The Suzhou jade industry at the Qianlong court), in Nancy Berliner (ed.), *Beyond the Screen: Chinese Furniture of the 16th and 17th Centuries*, Boston, MA: Museum of Fine Arts, pp. 31–75.

Wang Yuhu (ed.) (1957) *Qin Jin nongyan* (Agricultural precepts of Qin and Jin), Beijing: Zhonghua Press.

—— (1979) *Zhongguo nongxue shulu* (Bibliography of Chinese agronomy), Beijing: Nongye Press.

—— (1981, ms. completed 1966) *Wang Zhen Nongshu* (The Agricultural Treatise of Wang Zhen), Beijing: Nongye Press.

Wang Zhongmin (revised He Zhaowu) (1981) *Xu Guangqi*, Shanghai: Renmin Press.

Wang Zhongmin (ed.) (1984 [1963]) *Xu Guangqi ji* (Collected works of Xu Guangqi), 2 vols, Shanghai: Shanghai guji chubanshe.

Warnier, Jean-Pierre (2009) 'Technology as efficacious action on objects . . . and subjects', *Journal of Material Culture* 14, 4: 459–471.

Watabe Takeshi (1986) 'Chūgoku nōsho *Kōshikido* no ryūden to sono eikyō ni tsuite' (Studies about the spread and influences of 'Pictures of Tilling and Weaving'), *Bulletin of the Faculty of Letters, Tokai University* 46: 1–37.

Watabe Takeshi, Huo Wei and Christian Daniels (2003) *Shisen no dentō bunka to seikatsu gijutsu* (Everyday technology and traditional culture in Sichuan), Tokyo: Tokyo Foreign Languages University Press.

Watson, James L. (1982) 'Of flesh and bones: the management of death pollution in Cantonese society', in Maurice Bloch and Jonathan Parry (eds), *Death and the Regeneration of Life*, Cambridge: Cambridge University Press, pp. 155–186.

—— (1988) 'The structure of Chinese funerary rites: elementary forms, ritual sequence, and the primacy of performance', in James L. Watson and Evelyn S. Rawski (eds), *Death Ritual in Late Imperial and Modern China*, Berkeley, CA: University of California Press, pp. 3–19.

Watson, James L. and Evelyn S. Rawski (eds) (1988) *Death Ritual in Late Imperial and Modern China*, Berkeley, CA: University of California Press.

Watson, Rubie S. (1985) *Inequality Among Brothers: Class and Kinship in South China*, Cambridge: Cambridge University Press.

Watson, Rubie S. and Patricia B. Ebrey (eds) (1991) *Marriage and Inequality in Chinese Society*, Berkeley, CA: University of California Press.

Weiner, Annette B. (1992) *Inalienable Possessions: The Paradox of Keeping-While-Giving*, Berkeley, CA: California University Press.

Weiner, Annette B. and Jane Schneider (eds) (1989) *Cloth and Human Experience*, Washington, DC: Smithsonian Institution Press.

Wen Zhenheng (1996) (tr. Huanjing Xiu Maske) 'One Ming gentleman's notions on room order: selections from *Zhang Wu Zhi* – A Treatise on Superfluous Things', in Berliner Dagmar Schäfer (ed.), *Cultures of Knowledge: Technology in Chinese History*, Leiden: Brill., pp. 85–88.

White, Lynn (ed.) (1984) 'Symposium on Joseph Needham's *Science and Civilisation in China*', *Isis* 75: 715–725.

Will, Pierre-Étienne (1990) (trs. Elborg Foster) *Bureaucracy and Famine in Eighteenth-Century China*, Palo Alto, CA: Stanford University Press.

—— (1994) 'Développement quantitatif et développement qualitatif en Chine a la fin de l'époque impériale', *Annales Histoire, Sciences Sociales* 49, 4: 863–902.

Will, Pierre-Étienne and R. Bin Wong (1991) (with James Lee, contributions by Jean Oi, Peter Perdue) *Nourish the People: The State Civilian Granary System in China, 1650–1850*, Ann Arbor, MI: Center for Chinese Studies, University of Michigan.

Williams, Raymond (1973) *The Country and the City*, London: Fontana.

Wilson, Peter J. (1988) *The Domestication of the Human Species*, New Haven, CT: Yale University Press.

Wilson, Verity (1986) *Chinese Dress*, London: Victoria and Albert Museum.

Winchester, Simon (2008) *The Man Who Loved China: The Fantastic Story of the Eccentric Scientist Who Unlocked the Mysteries of the Middle Kingdom*, New York: Harper Collins.

Wittfogel, Karl (1957) *Oriental Despotism: A Comparative Study of Total Power*, New Haven, CT: Yale University Press.

Wolf, Arthur P. (2001) 'Is there evidence of birth control in Late Imperial China?', *Population and Development Review* 27, 1: 133–154.

Wong, R. Bin (1997) *China Transformed: Historical Change and the Limits of European Experience*, Ithaca, NY and London: Cornell University Press.

—— (2002) 'The search for European differences and domination in the early modern world: a view from Asia', *The American Historical Review* 107, 2: 447–469.

Wright, Arthur F. (1979) 'The Sui dynasty (581–617)', in Denis Twitchett and John K. Fairbank (eds), *The Cambridge History of China*, Volume 3, *Sui and T'ang China, 589–906*, Cambridge: Cambridge University Press, pp. 48–149.

Wright, Tim (2007) 'An economic cycle in imperial China? Revisiting Robert Hartwell on iron and coal', *Journal of the Economic and Social History of the Orient* 50, 4: 398–423.

Wu Xiujie (2008) 'Men purchase, women use: coping with domestic electrical applicances in rural China', in Francesca Bray (ed.), *East Asian Science, Technology and Society: An International Journal*, special issue, *Constructing Intimacy: Technology, Family and Gender in East Asia*, 2, 2: 211–234.

—— (2010) *Reproducing Women: Medicine, Metaphor and Childbirth in Late Imperial China*, Berkeley, CA: University of California Press., pp. 211–234.

Wu Yi-Li (2002) 'Ghost fetuses, false pregnancies, and the parameters of medical uncertainty in classical Chinese gynecology', *Nan Nü* 4, 2: 170–206.

Xia Weiying (ed.) (1956) *Lüshi chunqiu shangnong deng sipian jiaoshi* (The four chapters in the *Lüshi chunqiu* on agriculture emended and explained), Beijing: Zhonghua.

Xue Yong (2005) '"Treasure nightsoil as if it were gold": economic and ecological links between urban and rural areas in late imperial Jiangnan', *Late Imperial China* 26, 1: 41–71.

—— (ed.) (1969) *Tenkō kaibutsu* (*Tiangong kaiwu*), Tokyo: Heibonsha.

Zeitlin, Judith T. (2007) 'The literary fashioning of medical authority: a study of Sun Yikui's case histories', in Charlotte Furth, Judith T. Zeitlin, and Ping-chen Hsiung (eds),

Thinking with Cases: Specialist Knowledge in Chinese Cultural History, Honolulu, HI: University of Hawaii Press, pp. 169–202.

Zeng Xiongsheng (1998) 'Huang-lu rice in Chinese history', *Agricultural Archaeology* 3: 292–311. Translated/interpreted by W. Tsao, Ph.D. 10 October 2001, edited by B. Gordon, available at: http://http-server.carleton.ca/~bgordon/Rice/papers/zeng98.htm (accessed 7 August 2012).

—— (2008) *Zhongguo nongxue shi* (A history of Chinese agronomy), Fujian: Renmin Press.

Zhang Chi and Hsiao-chun Hung (2010) 'The emergence of agriculture in southern China', *Antiquity* 84: 11–25.

Zhang Qiong (2010) 'Huangquan yu jiyi: Qingdai nei zhiranju kaocha' (Imperial power and the crafts: an investigation of the Weaving and Dyeing Bureau during the Qing dynasty'), in Dagmar Schäfer (ed.), *Cultures of Knowledge: Technology in Chinese History*, Leiden: Brill., pp. 77–122.

Zhang Zhongge (1995) '*Zhongguo gudai gengzhi tu* xu' (Foreword to *Zhongguo gudai gengzhi tu*), in Chaosheng Wang (ed.), *Zhongguo gudai gengzhi tu* (Farming and weaving pictures in ancient China), Beijing: China Agriculture Press, pp. 1–4.

Zhao Zhongwei (2002) 'Fertility control in China's past', *Population and Development Review* 28, 4: 751–757.

Zilsel, Edgar (eds Diederick Raven, Wolfgang Krohn and Robert S. Cohen) (2000) *The Social Origins of Modern Science*, Dordrecht and Boston, MA: Kluwer Academic Publishers

INDEX